A Theory of Ecological Justice

As a result of human activities, many organisms on Earth face serious and worsening threats to their continued existence. This is usually regarded as a matter of concern because maintaining a healthy non-human environment affects the well-being of humans. *A Theory of Ecological Justice* adopts a very different approach, defending in detail the claim that all organisms, sentient and non-sentient, have a claim in justice to a fair share of the planet's environmental resources.

Baxter puts forward a detailed argument for accepting that all organisms count, morally speaking, and not simply those which are sentient or which human beings like or find useful. He explores the nature of justice claims as applied to organisms of various degrees of complexity and connects the concept of ecological justice to standard liberal theories of justice, such as those found in the work of Rawls and Barry. Finally, he explores the possibilities for achieving ecological justice within current national and international institutional arrangements for biodiversity protection. He also seeks to vindicate a universalist approach to moral thinking against those, postmodernists and others, who favour a contextualist view.

This book makes a thoroughly developed, ground-breaking case for ecological justice. It is an essential read for everyone interested in environmental politics and ethics.

Brian Baxter is a Senior Lecturer in Politics at the University of Dundee, where he teaches environmental politics and political philosophy. He is also the author of *Ecologism: an introduction* (1999) and co-editor of *Europe, Globalization and Sustainable Development* (2004).

Environmental Politics/Routledge Research in Environmental Politics

Edited by Matthew Paterson
Keele University

and

Graham Smith
University of Southampton

Over recent years environmental politics has moved from being a peripheral interest to a central concern within the discipline of politics. This series aims to reinforce this trend through the publication of books that investigate the nature of contemporary environmental politics and to show the centrality of environmental politics to the study of politics per se. The series understands politics in a broad sense, and books will focus on mainstream issues such as the policy process and new social movements, as well as emerging areas such as cultural politics and political economy. Books in the series will analyse contemporary political practices with regard to the environment and/or explore possible future directions for the 'greening' of contemporary politics. The series will not only be of interest to academics and students working in the environmental field, but will also demand to be read within the broader discipline.

The series consists of two strands:

Environmental Politics addresses the needs of students and teachers, and the titles will be published in paperback and hardback. Titles include:

Global Warming and Global Politics
Matthew Paterson

Politics and the Environment
James Connelly and Graham Smith

International Relations Theory and Ecological Thought
Towards synthesis
Edited by Eric Laffferière and Peter Stoett

Planning Sustainability
Edited by Michael Kenny and James Meadowcroft

Deliberative Democracy and the Environment
Graham Smith

Routledge Research in Environmental Politics presents innovative new research intended for high-level specialist readership. These titles are published in hardback only and include:

A Theory of Ecological Justice

Brian Baxter

 Routledge
Taylor & Francis Group

LONDON AND NEW YORK

First published 2005
by Routledge

2 Park Square, Milton Park, Abingdon, Oxfordshire OX14 4RN

Simultaneously published in the USA and Canada
by Routledge

711 Third Avenue, New York, NY 10017

First issued in paperback 2014

Routledge is an imprint of the Taylor & Francis Group, an informa company

© 2005 Brian Baxter

Typeset in Baskerville by
BOOK NOW Ltd

British Library Cataloguing in Publication Data
A catalogue record for this book is available from the British Library

Library of Congress Cataloging in Publication Data
Baxter, Brian, 1949–
    A theory of ecological justice/Brian Baxter.
        p. cm.
    Includes bibliographical references and index.
    1. Environmental ethics. 2. Conservation of natural resources. I.
Title.
    GE42.B39 2004
    179′.1–dc22   2004003907

ISBN 978-0-415-31139-7 (hbk)
ISBN 978-0-415-75854-3 (pbk)

To Lynn

Contents

x *Contents*

PART IV
Institutional arrangements for ecological justice 153

Acknowledgements

I have presented some of the ideas in this book at various places in recent years, and have received beneficial comments at all of them. An early version of Chapter 8 was presented to the politics research seminar of Keele University, where Andy Dobson and his students gave me valuable feedback. I gratefully acknowledge the permission of Frank Cass, publishers of *Environmental Politics*, and the editors of that journal to use a version of my arguments in Chapter 8 which they published in 2000 (vol. 9, pp. 43–64) as 'Ecological justice and justice as impartiality'. The published paper benefited greatly from the comments of John Horton and another, anonymous, reviewer.

A first version of the arguments which are to be found in Chapters 4 to 6 and Chapters 9 and 10 was presented to a meeting of the Interdisciplinary Research Network on Environment and Society (IRNES) in 2001 at the University of Keele, where John Barry and the other participants provided searching responses. Marcel Wissenburg very kindly gave me a detailed set of critical comments on this paper, to which I have done my best to respond.

Chapter 2 started life in 2002 as a paper to the Unit for Social and Environmental Research at the University of Abertay, where the target of many of its criticisms, Mick Smith, provided vigorous but courteous replies to my points. He and I have discussed the matters therein on several occasions and I believe that what I have to say on the issue of social constructivism has become clearer as a result.

A version of Chapters 11 and 12 was presented to the second biennial conference of the European Consortium for Political Research (ECPR) at Marburg, Germany, in 2003 as part of a panel which I organized on environmental and ecological justice. Conversations with, and the papers of, the other participants on that panel, namely Derek Bell, Simon Hailwood and David Schlosberg, have been of great benefit to my discussion of liberalism and ecological justice in Chapter 7.

Andy Dobson and Tim Hayward very kindly agreed to find time within their busy schedules to read the first draft, giving me valuable critical responses to which I have endeavoured to respond.

None of the above-named people who have been so generous with their time and comments bear any responsibility, of course, for what appears in the following pages.

I am grateful to my department, and its head, Norrie MacQueen, for allowing me to arrange my teaching duties in session 2003–4 so as to permit me to complete the book. Heidi Bagtazo and Grace McInnes have been very helpful and efficient editors at Routledge, and Carol Benoit-Ngassam and Susan Malloch have provided exemplary secretarial support within my department. Thanks, too, to our friends Colin and Ursula Doherty for providing me with a computer during a crucial writing phase in the summer of 2003.

Finally, I have dedicated this book to my wife, Lynn, with limitless gratitude both for her invaluable comments on the text and for her love and encouragement throughout its writing and throughout our life together.

1   The concept of ecological justice

In a three-part series entitled *The State of the Planet*, broadcast on BBC television in November 2000, Sir David Attenborough, the person associated, at least in British minds, with nature and the environment in all its forms, provided a lucid and alarming survey of the main ways in which human beings are currently destroying many of the plants and animals around them, driving many towards extinction. During the second episode, in the course of considering the reasons for the human-caused extinction of a species of snail unique to Hawaii, he addressed the issue of whether it mattered that one small species of snail should become extinct, especially as apparently no other ecological damage had resulted, as far as we know. His reply to this question was as follows: 'Surely it is sad indeed that our descendants should inherit a natural world that is more impoverished than the one we inherited?' (BBC 2000a). This sort of rather wistful response to the extinction of other species is often encountered. In voicing it David Attenborough was simply expressing a prevalent view, even among those who regard themselves as concerned for the natural environment. In further identifying as the injured party 'our descendants', he also adopted a standpoint which is widespread, even among those most concerned about the prospects of large-scale human-caused extinctions of other species. The possibility of there being any wrong done to the species of snail, or to the individual members of the species, receives no mention at all, not even to be dismissed. Apparently it is only the possible losses to actual and future human beings, whether aesthetic, cultural, scientific, medical, economic, recreational and so forth, that count.

One way to understand the point of this book is to see it as seeking to establish the wholly inadequate character of these responses. If the arguments presented below are correct we should rather say that human beings would have done something grossly unjust if they were to perpetrate the extinctions of other life-forms, as envisaged by Attenborough and many other well-informed commentators, when they could take steps to avoid doing so without serious harm to human life. This specific injustice, further, will have been done to the creatures themselves, not to human beings.

In order to flesh out this claim a little further it will be as well to begin with an outline of how the approach of this book towards human beings' relationships with non-human life differs from that of other traditions and theories. The first point to make is that human beings have always had at least some concern to treat at least some of the other organisms with which they share the planet with what they regard as due respect (see Perlo 2003; Sylvan and Bennett 1994: 29–30). The basis of this respect differs across time and place. Often it is bound up with some religious belief which establishes obligation-grounding relationships between all human beings, or at least the believer's own group, and other forms of life – either in general, or in specific categories. Sometimes this has involved the idea that other creatures share an important aspect of their inner selves in common with human beings. Sometimes a myth of a common ancestor does the job, sometimes a doctrine of reincarnation. Often various elements of such thoughts will all play a part. Often the idea that certain creatures – plants or animals – are sacred is important here. Other forms of religious-based concern for the non-human world involve a more indirect form of responsibility for it – a responsibility held directly to the benevolent creator for providing it for human use. In such patterns of thinking the idea of a duty of human stewardship towards other life-forms emerges as the guiding conception (see Attfield 1991; Daly and Cobb 1990).

In an age when religious ideas lose some of their power and plausibility, secular versions of at least some of these conceptions begin to emerge. For many modern people, the idea of the kinship between all life-forms, as revealed by Darwin's and Wallace's theory of evolution, plays an important moral function in sustaining a sense of human obligation towards other life-forms. The concept of stewardship has also been formulated in non-religious terms. In an age still dominated by the utilitarianism which emerged in its clearest form in the eighteenth and nineteenth centuries in Europe, it was the capacity of some organisms to feel pleasure and pain which gained them admission to the area of moral concern. Animal rights advocates often focus upon this element of other organisms' capacities to make the moral case against factory farming, scientific experiment using animals, and so forth.

In all these instances, whether religious or secular, we can see that human beings have for long found reason to adopt an attitude of concern towards at least some aspect of the non-human world, and to see it as a requirement of morality, in some form, to restrict their behaviour in some ways. Failure to do so results in impiety, sacrilege, inhumanity or cruelty.

None of these traditional ways of conceiving of the morality of human relationships with non-human organisms is the same as behaviour-influencing considerations drawn from the purely prudential concern to maintain a healthy non-human environment for the benefit of existing or future human beings, or groups thereof. A great deal of the modern Western concern for the environment is drawn from this source. We human beings derive all of our resources, directly or indirectly, from the natural world, and this fact has played a great and significant part in the modern environmental movement's effort to articulate our continued interconnectedness with the biosphere of the planet, and to point to the various serious ways in which we appear to be harming it.

Such concerns have not been without effect in the world of politics, law and international relations. We now inhabit a world in which serious steps have been, and are being, taken to preserve species and habitats from human-caused damage.

These have involved whole rafts of legislation to protect, in some way, endangered species within local and state jurisdictions, across state boundaries and in the open access areas, such as the open sea, beyond the reach of any one state's jurisdiction. By one recent reckoning there now exist over 300 multilateral agreements to protect the environment. Of these, nearly one-third are at least in part designed to 'protect biodiversity', which is the phrase which has been coined to refer to the protection of endangered species. Although most of these are regional, thirteen have a global coverage (McGraw 2002: 9–10). There now exists a growing body of ecologists and conservation managers seeking to apply our scientific under- standing of ecosystems to the maintenance of protected areas within which endangered species may continue to survive, and to foster their survival beyond protected areas by advising human economic actors how to operate in such a way as to remove the threats which their actions might otherwise pose for other life-forms.

To sum up, then, one can say that it is by no means a new or unusual thought that human beings have some moral responsibilities towards at least some parts of the non-human world. It is equally apparent that the idea that it is in the interest of human beings to maintain a healthy environment for themselves and their descendants is one which readily finds many supporters. Finally, these moti- vations have led human states to legislate and organize to provide some measure of protection for other life-forms, and to employ professional environmental care- takers to implement this protection.

From the point of view defended in this book these are all welcome phenomena. It is, of course, entirely possible to raise a sceptical voice against many of the forms of concern just outlined. The religious-based moral conceptions are open to criticism from the secular standpoint as part of the general critique of religious thought. Many from the religious side will see the secular approaches to environ- mental protection as doomed to failure unless human beings manage to recover a sense of the sacred. One may well ask how effective in practice are the legislative efforts, apparently so extensive, to protect the environment in the face of powerful economic, political and ideological forces operating in the opposite direction. Plainly, then, none of these phenomena can be treated as wholly unproblematic. However, their existence does at least help with the project of this book by showing that the very idea of seeking to protect the continued existence of non- human creatures on moral grounds is one which has a long history of human acceptance. Furthermore, the existence of institutions at all levels of human political organization – legislation, treaties, conventions, conservation organiza- tions and managers – will play a very important role later in the book when we come to consider what institutional arrangements will be necessary to secure ecological justice.

However, the premise with which this book begins is that none of these existing ways, whether moral or purely prudential, of thinking about how human beings should conduct themselves towards non-human beings is fully satisfactory. The moral forms of thought either do not cover all relevant non-human life-forms, or cover them for the wrong reasons, morally speaking. For example, they restrict concern to some category of the sacred or to the sentient, or, when all life-forms are covered, as in the stewardship traditions, they operate with notion too vague to be very helpful. The purely prudential basis for producing such protection, based on the idea that it is in human self-interest to afford it, sounds tough-minded enough to appeal 'in the real world'. However, it has always been, and may well remain, a matter of contention between ecologists how much species variety a 'healthy' environment needs, whether there are any 'redundant species' from the point of view of ecosystem health, and so on. Hence there will probably always be plenty of 'wriggle room' to allow decision-makers to permit human-caused exterminations to proceed on this basis. As Andrew Dobson has argued, this deficiency is apparent in the approach of the Brundtland Report to sustainable development (Dobson 2000: 59). That contains a purely instrumental (what I am here calling a 'prudential') defence of biodiversity which is not to be sneezed at, but which contains a worrying element of contingency in the connection between human self-interest and the survival of other species. If you begin, as many do, with the intuition that there is something intrinsically wrong with human-caused extinctions of other species, then this prudential approach will cause grave disquiet.

The view propounded in this book is defended on the basis that it is the best way of articulating and defending the intuition mentioned in the last paragraph, whatever practical conclusions follow from it, but also on the basis that it is needed to push the protection of endangered life-forms high enough up the scale of human moral concern. Only if we move beyond the idea that it would merely be 'sad' if large numbers of other life-forms are exterminated and become convinced by the idea that a serious injustice will have been done will we be likely to take the avoidance of that outcome with any real seriousness. Further, the concept of justice to be employed in this argument is a specific one – that of distributive justice – rather than the more general one which employs the term as almost synonymous with 'morally right actions'. We must do right by other life-forms, but in a precise kind of way, namely by recognizing their claim to a fair share of the environmental resources which all life-forms need to survive and to flourish.

Of course, a moral argument of the kind offered in this book is unlikely all on its own to convince the unpersuaded. Notwithstanding the points made above about the prevalence across time and space of human beings' receptivity to the idea of moral obligations of some sort towards the non-human, the present situation is one which requires that receptivity to be deepened, extended and located within a basic understanding of what 6 billion and more human beings are in the course of doing to the planet. Although the claims made on this issue are the subject of ongoing controversy, one might expect that the prima facie importance of the issues would lead to at least an interest on the part of an informed public. Many, of course, are so interested. Many, however, are not, or not to any significant degree. The following passage from a recent volume devoted to conservation management makes sobering reading in this regard:

> The linkage of ecology to the every day concerns of people is highlighted by a survey done by the Consultative Group on Biological Diversity. That group

engaged a professional survey firm to assess the views of voters who were neither committed to nor hostile toward the environment . . . This politically engaged but moderate public had little deep concern with environmental issues. Introducing other concerns, say crime or jobs, easily dislodged environmental concerns from their pride of place. The term "biodiversity," which we suspect scientists view as a triumph of public relations savvy, seems to the public as a divisive and somehow negative idea . . . Contemporary human-caused extinctions are confused with the natural extinctions of the past. Extinctions are seen as something that have always happened . . . And . . . people do not see what species richness or degradation of ecosystem function have to do with them.

(Pickett *et al.* 1997: 398)

This is echoed in a recent comment on the public invisibility of the major international agreement which has the closest connection with the aims of this book – the 1992 Convention on Biological Diversity (CBD):

The CBD reached its peak in popularity when the US announced it would not sign it. Since that time, the Convention has received negligible coverage in the mainstream media – especially when compared to its ozone and climate change counterparts. If the CBD is indeed viewed as both less popular and less prestigious than these other agreements, it is in part due to the nature of the issue area itself. Both the breadth and depth of biodiversity make it difficult to define a clear *problematique*.

(McGraw 2002: 26)

Plainly, a great deal of work remains to be done to produce the basic ecological understanding necessary among such groups of citizens in all the states of the world for a grasp of what the moral issues raised in this book and elsewhere are supposed to be. And moral theorists are also uneasily aware that moral argument, even when successful, is always problematic. People do not welcome being preached at or being made to feel guilty, however just the cause, and are probably most willing to accept moral prescriptions when they are firmly located in a religious world view which they are prepared to accept.

However, if an issue is a moral one then it does no good in the long run to pretend otherwise. We certainly need to appeal to the self-interest of human beings as well as to their moral sense in order to secure the changes in human behaviour necessary to achieve a world as rich in species as the current one. The concept of self-interest will need to be an enriched one, such as that eloquently developed by Tim Hayward (Hayward 1998), appealing to the intellectual, aesthetic, recreational and spiritual benefits which it is possible for at least many human beings to derive from ongoing encounters with the natural world, as well as to the material and economic advantages it provides. However, this book is written from the conviction that there is an important moral relationship here which needs to be stated and defended, and that an acceptance of it will significantly contribute to a better position for the goal of protection of endangered life-forms on the agenda of human action.

In order to explore this position a bit further we will need to explain the concept of justice with which we are operating here, and to do that we will need to begin with a brief survey of the ways in which the concept of justice has been developing in recent decades.

The idea of justice

For political philosophers working in the tradition of analytical philosophy the issue of justice, especially social or distributive justice, has been at the centre of attention ever since John Rawls put it there in the early 1970s (Rawls 1972). Much ingenuity has been devoted in the last thirty years to the development and critique of Rawls's ideas and the offering of alternatives. For most of that time mainstream theorizing has worked with a distinctive view of what justice is about. This is that it essentially applies within a single society of contemporaneous human beings seeking to reach agreement on the basic rules to govern their lives together. It is only quite recently that this focus upon the domestic political order of a single society has been challenged.

As a result, issues of distributive justice, as Rawls himself latterly agreed (Rawls 1993: 245), are now frequently said to encompass two further broad groups of human beings: contemporary aliens, under the heading of 'international distributive justice', and people who have not yet been born, under the heading of 'justice to future generations' (see, for example, Beitz 1979; Sikora and Barry 1996). This is the result, to put the matter at its most general, of our increasing realization that human beings have important impacts upon each other's well-being even when they do not inhabit the same society or historical period. Economic, cultural and environmental processes are the vehicles for these kinds of impact, of course, and they appear to raise matters which are indisputably matters of justice – what rights people have, how they may properly be treated and to what resources they are entitled.

Among the latter issues, that of environmental sustainability, which forms a large part of the problem of what we are morally required to bequeath to future generations, has emerged as the focus of much debate (see, for example, Dobson 1999). This forms part of a more general set of issues concerning the just distribution of environmental 'goods', such as agricultural land, clean water, and mineral resources, and 'bads', such as landfill sites and toxic waste disposal plants. This set of issues – how environmental goods and bads are to be distributed among human beings, within and across societies at any one time, and between generations across time – has recently received the label 'environmental justice' (see Low and Gleeson 1998). In the USA there has even developed an 'environmental justice' movement which has been inspired by just such concerns, especially with the combating of environmental racism – the distribution of environmental bads in ways which unfairly impinge upon particular racial groupings (see Szasz 1994; Dowie 1995: ch. 6). Clearly, then, this dimension of environmental justice cuts across the issues of intra-societal, international and intergenerational justice just identified.

In addition to all of these ways in which the issues of distributive justice between human beings have been made more complex, a further, more controversial, dimension has been introduced into the debate specifically by green theorists. This is the dimension of human–non-human relations. At a minimum, it involves the claim that at least some aspects of non-human nature are worthy of moral consideration – 'possess moral considerability', to use the term favoured in the literature (Goodpaster 1978). The idea is that non-human organisms, although not themselves moral agents, may intelligibly be said to have claims upon the actions of moral agents. On this planet, of course, the only moral agents we know of are human beings.

These claims more ambitiously involve postulating the prima facie right of organisms to continue to exist within the habitats required to sustain their existence so as to reproduce their own kind and maintain their existence as a species. It is recognized by those who put it forward that this prima facie right may be overridden in certain specified circumstances. The extermination of the smallpox virus is an obvious case in point – an extreme case of the conflict of interests between one organism and others which most would resolve in favour of the others.

But the idea behind this conception of human–non-human moral relations is that non-human organisms have a claim in justice against moral agents not to deprive them without good moral reason of the environmental basis of their continued existence and ability to reproduce themselves. This is where the idea of 'ecological justice' enters the picture. We should note in passing that Rawls balked at the idea that the relations between human beings and non-human organisms could properly be brought under the heading of 'justice' even though he was equally clear that human beings have some moral responsibilities towards non-human life-forms (Rawls 1999: 448). We will need to consider these qualms at a later stage.

The term 'ecological justice' is one which seems first to have been coined by Low and Gleeson (1998). They introduce it as follows:

> The struggle for justice as it is shaped by the politics of the environment . . . has two relational aspects: the justice of the distribution of environments among peoples, and the justice of the relations between humans and the rest of the natural world. We term these aspects of justice: *environmental* justice and *ecological* justice. They are really two aspects of the same relationship.
>
> (Low and Gleeson 1998: 2)

The term is one for which there has been a need in environmental political theorizing. Even if you believe, as many do, that the relations between human beings and the non-human world cannot embody requirements of justice, it is useful to have a label to designate succinctly what you are rejecting (compare Hayek's and Nozick's rejection of 'social/distributive justice' (Hayek 1960: 93,

385; Nozick 1974: chs 7, 8). However, the underlying assumption of this book is that the relations between human beings and non-human nature do encompass issues of justice, specifically distributive justice, hence the term is to be welcomed as usefully delimiting an area of the theory of justice which needs elaboration and defence.

The concept of ecological justice, however, does not simply identify a theoretical lacuna in the theory of justice. Rather, as mentioned earlier, it is urgently needed to elaborate an adequate practical moral response to the pressure which *Homo sapiens* has recently been exerting upon the other species of this planet. This book is based on the assumption that this pressure, as many have argued, threatens a rate of extinction of many of those species to rival that of the great extinction episodes in previous eras of the planet's history (see Leakey and Lewin 1996). It is because of this phenomenon that it now seems to some to be necessary to argue in the strongest moral terms which we have available in our Western, and perhaps global, moral vocabulary that other species have rights to what is necessary in the way of environmental resources in order to survive, flourish and perpetuate their kind. However, even if this empirical premise should turn out to be mistaken, the moral case will remain untouched – ecological justice will remain a requirement governing human action even if few other species are being put under extermination pressure by human beings.

Plan of this book

We have seen that many theorists, as well as many long-standing cultural traditions, are willing to accept that human beings have some moral responsibility towards at least some non-human organisms. The aim of this book, however, is ambitious in at least two respects. Firstly, it aims to show that there is a strong case for extending the very specific moral category of distributive justice to non-human life-forms. Secondly, it seeks to show that this concept covers all forms of life, and not just some. In order to make this comprehensive case for ecological justice we will have to argue firstly for the claim that all non-human organisms are morally considerable. This is because some very important theorists of the moral considerability of the non-human have not been willing to extend this status to all non-human organisms. The most favoured limitation is that of sentience.

Accordingly, we will have to spend time establishing, against such theorists, that all organisms, even the 'merely living', possess moral considerability. We will first discuss the arguments of David DeGrazia and Peter Singer, which demonstrate that non-human animals do possess significant moral status. We will then show that their restriction of this status to the subcategory of sentient organisms is unwarranted. This will lead to a consideration of two recent theories of moral status which accord the non-sentient some moral status, but on rather different theoretical bases. These are the theories of Mary Anne Warren and Jon Wetlesen. Consideration of their theories will bring into focus issues of meta-ethics which will require resolution in the course of the development of a theory of ecological justice.

However, even if we do develop these arguments to their fullest extent, by showing that all organisms possess some degree of moral status, this will not in itself show that such organisms are the proper recipients of distributive justice, which, as we have noted, is a very specific moral category. As we have seen, Rawls, among others, refused to accept that non-human organisms could be the proper recipients of distributional justice. We will, therefore, have to spend some time establishing the following claims:

1   that all non-humans, sentient and non-sentient, are members of the community of justice;
2   that all members of the community of justice are proper recipients of distributive justice with respect to environmental goods and bads – that is, to ecological justice.

We will consider the objections in principle to extending the concept of distributive justice to the non-human, and show that these are unsustainable. We will then consider the dominant contemporary analysis of distributive justice, as initiated by Rawls and developed by Brian Barry in his theory of 'justice as impartiality', in an effort to show how the basic, constitutionally entrenched institutions of a liberal society can encompass the requirements of ecological justice. We will have to consider the objection to the idea of such constitutional entrenchment that it is in some way anti-democratic or authoritarian (Bell 2003). This will inevitably involve some consideration of a recently much-debated question, which is how far the dominant ideology of liberalism can encompass the moral considerability of the non-human in general, and the concept of ecological justice in particular, and how far an ecologically just society must be 'post-liberal' (Barry, J., 1999: 92).

Having established the case for ecological justice against its detractors, and shown how it can be related to the most developed theories of distributive justice within the liberal tradition, we will consider the specific requirements of ecological justice. This will first encompass the non-sentient forms of organic life – the 'merely living' world of bacteria, plants and so on. We will then consider the world of the 'higher' sentient organisms, culminating in our closest relatives, the great apes. In each case it will be necessary to determine who are the appropriate recipients of ecological justice – individuals, populations or species – and to what, precisely, ecological justice establishes their prima facie claim. It will also be important to work out how clashes of claims are to be adjudicated, and how ecological justice relates to distributive justice between human beings.

The final section of the book will be taken up with an examination of the institutional arrangements necessary, both within and between states, for the securing of ecological justice. In this regard the approach adopted will be one which eschews the utopian strategy, particularly prevalent among environmental theorists influenced by the anti-state arguments of the anarchist tradition, of seeking to draw up an ideal blueprint of the environmentally sustainable society in which the state is replaced by some more favoured political structures. Rather, the developments in the actual world, such as the creation of international regimes for environmental protection in general and biodiversity protection in particular, and of conservation management strategies, will be closely examined with an eye to the possibilities which they permit for adaptation to the purposes of ecological justice. Some reasons for a strongly qualified optimism will be found as the result of this investigation.

It will be apparent from this outline that it is possible for there to be a less comprehensive theory of ecological justice than the one defended in this book. If the arguments for including non-sentients under justice fail, or if the arguments for extending moral considerability in general to the non-human fail, this would still leave open the possibility that some non-humans – all, or at least some, of the sentient ones – could be covered by justice requirements. This book is written, however, on the assumption that a moral case which covers all life-forms is preferable if it can be attained. This means that we should at least begin by trying as hard as possible to generate such a case, even if we have to fall back on a more limited aim should valid arguments not be achievable.

However, before embarking on these endeavours it is necessary to devote some time to the elucidation and defence of the universalist approach to moral debate which is adopted here. For there are many contemporary environmental theorists, among them eloquent defenders of the moral standing of non-person organisms, who are deeply critical of the idea that it is possible to generate an ethical position in the manner attempted in this book. This involves articulating a universalist moral position, addressed to all moral agents, and assuming that arguments can be propounded which are capable of influencing any moral agent who cares to attend to them.

This approach to moral discussion has some strong detractors nowadays. It has been characterized as part of a discredited Enlightenment, modernist project based on a historically and culturally naïve view of how thought and action are related. It ignores the importance of social and historical context within which moral thought develops, and the important role of social constructivism for all fact–value combinations. More damagingly still, it has been charged with promulgating a concept of reason and morality which sustains the anti-human and anti-environment actions of self-serving, powerful, Western-based elites – capitalists, bureaucrats, self-proclaimed experts – and their lackeys in the world of science and technology.

Various parts of such a charge have become quite commonplace among radical environmental theorists. Many of the criticisms offered by such theorists are, in whole or part, valid. But we need to see that there is quite a lot which is invalid too. There are many contemporary environmental theorists (such as Rolston 1997) who would take issue with this anti-modernist, anti-universalist critique, and this book sides with them, even if it does not necessarily endorse some of the particular arguments which they have put forward. It is clear that it behoves anyone nowadays who attempts the kind of universalist approach used in this book to make some attempt to vindicate it against its searching and eloquent detractors. This is the task to which we now turn.

Part I

How to think about moral issues

Universalist versus contextualist approaches

2 The case for social constructivism considered

What is at issue in the universalist–contextualist dispute

The claim that this book puts forward is that the issue of distributive justice between species with respect to environmental resources is a meaningful and important one, and that a set of general considerations to guide moral persons in their deliberation on such matters may be elucidated. These considerations are intended to have perfectly general applicability, to be recognizable as reasonable by all moral persons who care to inspect them. This claim, therefore, supposes that it is possible for there to be something like a perfectly general moral language which persons from all cultural backgrounds can understand and deploy in the course of moral deliberation and argument.

In taking this view, I am nailing my colours to the mast of moral universalism, thereby joining with other moral theorists who have shown no particular interest in the putative issue of distributive justice between different species. For example, there are those who believe that issues of distributive justice between human beings from different cultures may similarly be discussed in a perfectly general way. In fact, as noted in the last chapter, the concept of international distributive justice is nowadays recognized as a prima facie defensible extension of the concept of distributive justice between members of the same community. Hence, in taking the universalist view, ecological justice theorists are to that extent not isolated in their conception of what is defensible in the field of moral theory.

Of course, the universalism of ecological justice may well have a different basis from that of other universalist discourses. For example, it can (and I think should) be given a basis in some form of naturalistic ethics, which seeks to vindicate the universalism of discourse about justice and other moral concepts in terms of a shared human nature, ultimately to be explained in evolutionary terms. Other forms of universalist discourse may be based on alternative general features of all human moral agents (or perhaps all moral agents *simpliciter*), such as rationality, considered as a set of abstract constraints upon thought. We should note, in passing, that these may not be exclusive categories – some ethical theorists have recently begun to argue that rationality can and should be given a naturalist explanation (see Preston 2002).

However, whatever the basis sought for moral universalism, it is clear that it is a conception which appears to many nowadays to be simply indefensible. Let us, therefore, examine how far these critics of moral universalism are correct in what they say, and whether the concerns which animate the defenders of the idea of ecological justice can be better pursued on the basis of some alternative, non-universalist, approach to an understanding of moral and other kinds of thought.

The approaches to the understanding of moral and all other forms of thought which radically eschew all forms of universalism are those which view all thought as necessarily context-dependent. Further, these approaches claim that the contexts in question are culturally constructed. In making this claim, however, they put forward a view which should be clearly distinguished from other kinds of theory which make an essential reference to context as the basis for understanding some aspect of the human predicament. For example, one theory takes the contexts in question to be objectively given by the natural circumstances in which various species, including the human species, find themselves. The environmental philosophy of Bryan Norton, for example, which emphasizes the importance of viewing human beings in the appropriate ecological context, as revealed by ecology and other biological sciences, is an example of such non-culturally constructed contextualism (Norton 1991: 237).

Another example comes from the area of educational theory, where the importance of the context within which learners develop (or 'construct') their knowledge is widely recognized as a vital part of the learning process. This insight has been applied to the development of a phenomenon of great importance to this book, namely biodiversity protection projects in which scientific knowledge and the knowledge of indigenous peoples have been brought into fruitful contact. This has involved recognition that the simple enunciation of what are held to be scientific truths about biodiversity and ecology rarely has the desired effect on the target audience. As Kelsey has put the point: 'Recognition that individuals are *active* agents in learning . . . demands that biodiversity education activities must be truly participatory. Furthermore, recognition that individuals construct know-ledge in specific contexts demands that biodiversity education activities must be planned and implemented *within* biodiversity projects' (Kelsey 2003: 390). This, too, is a view with which no serious issue needs to be taken, for it is unexception-able to claim that many, perhaps all, activities, learning among them, can only practicably take place when suitable contexts are provided, including ingredients based on the existing beliefs and competences of the learners.

These views are to be distinguished from the more radical position that all thought, moral and otherwise, is contextualized by an indefinite series of cultural constructs which may be incommensurable and thus not relatable to each other (MacIntyre 1988: 9). This is the guiding thought of all forms of postmodernism. Hence, our enquiry at this point can be construed as the attempt to bring the concept of ecological justice into the spotlight of the postmodern critique of all forms of universalism, especially of the modernist universalist discourse of progress, reason and universal rights.[1]

But there are various versions of postmodern theorizing, with some of the proponents of the importance of culturally constructed contexts rather unhappy to be labelled 'postmodernist'. One key point of disagreement within the postmodern camp, for example, is precisely over the extent to which different cultural constructs are to be regarded as incommensurable. One might take the view that the most radical forms of postmodernism regard such incommensurability as unavoidable, so that divergences between different contexts cannot be reconciled at some higher level, in a meta context which embraces the lower-order ones (see Rorty 1980: pt III). Such a position appears to embrace a thorough-going relativism which, for many, will appear to be ultimately incoherent.

But it might be possible to emphasize the centrality of culturally constructed contexts in the creation of modes of human thought without abandoning the possibility of intercultural dialogue which can, in the course of time, result in a (perhaps fairly rough-and-ready) convergence upon a single culturally constructed context that facilitates at least a large measure of interpersonal discussion and agreement on moral and other matters. The aim of the proponents of such postmodern views will then be primarily to set their faces against the imposition (as they see it) of abstractly generated moral positions (of the kind regarded as inherent in universalist discourse) by the representatives of a discredited modernity upon (sub)cultures and societies which draw their sustenance from outside the realms of modernism (see MacIntyre 1988: ch. 19).

The aim of this latter kind of postmodernist thought is thus to complexify the business of moral debate by casting it as a protracted and probably difficult series of dialogues between people approaching moral issues from a variety of cultural standpoints, groping towards forms of (perhaps temporary) agreement for particular purposes in hand and doing so on the only possible basis upon which it might succeed, namely by according each other's moral constructs respect and being highly self-conscious with respect to the cultural foundations of their own moral preconceptions (Rorty 1980: 315–22). The various forms of discourse ethics, such as that of Habermas (Habermas 1990), come to mind here. However, as is well known, this view is propounded by a self-proclaimed modernist, and bears at least some resemblance to those positions in the philosophy of natural science which seek to vindicate the general idea of progress in science with a due recognition of the importance of cultural context in the generation of scientific theories (see Kuhn 1970: 160–210). At this point, therefore, one may argue that we are really considering more complex versions of modernism, ones which have not really abandoned the idea of 'grand narrative', but rather make the narrative more fraught and tentative than those which inhabitants of modernist cultures have been used to.

There are other themes in the discussion of moral thought which feed into the issue of the importance of culturally constructed contexts. One of these, highlighted by some ecofeminist theorists, such as Karen Warren (Warren 2000), and by Mick Smith (Smith 2001) (who is one of those somewhat uncomfortable with the label 'postmodernist' to characterize his own 'ethics of place'), is the issue of how far moral thinking rests on purely rational foundations, and how far some form of emotional response to entities is central to moral behaviour with respect to them. Both of these theorists argue against what one might call the 'algorithmic' view of moral thought, according to which moral conclusions can be generated by the application of clear-cut moral principles to situations specifiable in clear and unambiguous moral terms. Both argue for the importance, instead, of caring as an emotional–attitudinal factor directed towards items regarded as possessing moral status.

Further, both reject the idea that some item's possession of moral status is a matter of its resemblance to some canonical moral beings, such as Western, rationalist-oriented human males, and emphasize the importance of the direct, emotion-based response to those things we care about morally, even when they bear little resemblance to us. In the case of Warren, there is a concomitant emphasis on the importance of narrative as the vitally important mode of thought in terms of which even alien others may be conceived of as appropriate objects of our moral caring (Warren 2000: 102–5). In the case of Smith, there is the emphasis on epiphany experiences, ones which often catch one unawares, in which some element, often of the non-human world, breaks through even the culturally constructed modes of understanding of the world, and 'speaks to us' directly in a visceral fashion (Smith 2001: 167).

The high quality of Smith's discussion makes it worth while to try to answer his arguments in some detail, in order to see how much of an inroad they make into the universalist approach favoured in this book, and how far the concerns of ecological justice may be found a place within his, and others', preferred version of an ethics of context.

Smith's case for social constructivism

Mick Smith's views are worth considering in some detail because he is seeking to defend the intrinsic value of non-human life-forms, and to argue for human beings to act in such as way as to respect that intrinsic value. His general aims are thus broadly the same as those of this book. His view of how to make the ethical case is, however, very different. Specifically, he diagnoses the main culprits in the current assault on the natural world as being those universalist forms of moral and scientific thought which form a central part of the European Enlightenment tradition. He aims to criticize that tradition by revealing its inextricable connection with the interests of certain powerful groups in the contemporary world and to vindicate an alternative, anarchist and contextualist view of values and beliefs which, in his view, represent the only basis for a form of human praxis that can find the appropriate forms of respect for the natural world.

Smith tells us that his aim is to effect a reconciliation between 'deep ecologists', by which he means those who argue for the intrinsic value of the natural world – many of whom, he tells us, have 'retreated into forms of philosophical objectivism and biological reductionism in order to defend nature's intrinsic value' (Smith 2001: 110) – and 'social constructivists', who are said to possess 'insights into the influences of different social circumstances in our conceptions of nature' (ibid.). The former, he tells us, have 'much to gain from constructivism, and can add a new dimension to constructivism's own critique of current ideologies' (ibid.).

To illustrate this idea Smith begins by surveying some of the diverse ways in which, both within and across cultures and times, different human groups and individuals have taken different views about the significance and value of one set of non-human organisms, namely trees (Smith 2001: 110–14). However, one might conclude from the examples he gives that this variety of responses, often from people who appear to inhabit the same society (for example, the conflicting views of North American loggers and environmental activists) cannot be explained in any very clear way in terms of 'different social circumstances'. All that can be said is that, with respect to any given set of ideas about the value of trees, often many people share them, and that their doing so may have to do with the defence of the economic interests which they believe themselves to have in common. But sometimes their acceptance of such values appears to be quite idiosyncratic – such as the example Smith cites from W. H. Hudson of the owner of a large estate who viewed his trees as possessing intelligent souls (ibid.: 113).

In neither case does it appear adequate to suggest that we can explain the fact that the individuals in question hold the views they do because of their 'cultural circumstances' rather than because they believe themselves to have good reason (often in the face of what is commonly believed in their culture) for the beliefs they hold. In any case, the example which Smith gives of the conflicting views of trees held by different Indian thinkers – the sociologists Gadgil and Guha view them as resources; Kerala untouchables view them as sacred – shows the difficulty of identifying the appropriate 'culture' which is doing the 'constructing'. Gadgil and Guha apparently think there is 'an imported neo-colonialist' (Smith 2001: 112) discourse operating among the indigenous 'ecosystem people', while Smith in turn sees Gadgil and Guha as possibly subject to 'the pervasive influence of a modernist Western ideology that emphasizes a particular "rational" and technological solution to so-called underdevelopment' (ibid.).

These kinds of claim, whether valid or not, highlight the difficulty of determining to which 'culture' a given instance of constructivism is to be attributed. Once we allow that people, whatever their social position or group membership, can be subject to influences from apparently very different societies and cultures from the one they are said to inhabit, then the concept of social constructivism becomes extremely slippery. It looks as if, given an individual's belief-system, we identify the appropriate culture which is said to have 'constructed' it as the culture within which it is reasonably common, whether or not we would otherwise identify the individual as a member of that culture. The connection is then made via the notion of 'influence'. How do we know that such influence has occurred in the individual's case? It is tempting at this point, but perhaps not unavoidable, to collapse the argument into a vicious circle by replying that the influence must have occurred, given that the individual holds the beliefs in question!

The more fundamental problem which lurks within the concept of social constructivism is that it undercuts the whole idea of people as holding their beliefs for reasons. Rather they are held to be influenced by cultural forces of which they are sometimes not even aware. In an even more radical formulation, it may even be claimed that concepts of 'reason' and 'rationality' are themselves socially constructed, so that what seems rational and a 'good' reason to members of one cultural formation will not seem so to someone from another formation. What this line of thought leads to is that people are diagnosed rather than reasoned with. Their views are pigeonholed as exemplars of an impersonal belief-system of which they are deemed to be the mere conduits. This is arguably in itself a dehumanizing move (although since the concept of what is 'human' is itself held to be socially constructed, this claim too invites diagnosis rather than argument).

But there are also fundamental and well-known, logical difficulties lurking here. To understand these we need to note first that there are two forms of social constructivism – a moderate and an extreme view. The more moderate view is that, while it is possible for people to hold their beliefs for reasons which can be assessed by all intelligent minds, they usually or often do not. Rather they are indoctrinated by their 'cultures' (supposing we can identify these without too much difficulty). The correct reply to this claim is that, even if it is true, it is irrelevant to the business of trying to determine which are the beliefs that could be held for good reasons, and which are not. It is clearly going to be this task which is logically and practically primary. Once we know what people have good reason to believe, we can hope to persuade them by citing those reasons. Only when they persist in rejecting them without good reason does it become relevant to look for the non-rational forces which may be operating in their minds. But until we know what are the good reasons we will be unable to identify what are non-rational forces, and thus be unable to take adequate steps to combat them.

There is, however, the more radical view to consider. This is that there is no non-socially constructed concept of reason available to us, and thus that there is no neutral concept of reason which would enable us to assess which socially constructed belief systems are rationally preferable to others. But this claim has a serious logical difficulty lurking within it. For it implies that not only can members of different cultural structures not have meaningful discussions with each other (for what each takes to be a good reason for holding a view will be different from that of their culturally alien interlocutor, unless there just happen to be overlaps), but even this statement of the situation will be devoid of general significance, for what is taken to be a good reason for holding it will also be culturally constructed, and thus purely idiosyncratic to a specific cultural formation.

Supporters of radical forms of social constructivism seem not to notice that their own social constructionist theory is created from within a cultural formation yet purports to make claims about all social constructions. It thus can have only non-universal significance (on the theory) yet purports to have universal significance. It thus contains a logical contradiction, visible from within the theory itself.

If, on the other hand, it is possible, even from within a cultural formation, to produce universally valid claims, then, of course, it does not seem impossible for any of us to escape in other ways from the confines of any cultural construction, albeit perhaps only with protracted effort. This then takes us back to the point made above in connection with the more moderate version of social constructivism, namely that the thesis of social constructivism is irrelevant to the assessment of beliefs. Social constructivism, therefore, seems to be either at best cautionary (make sure that you have reasons for your beliefs, for it is possible that you hold them as the result of social conditioning) or at worst a recipe for intellectual paralysis deriving from the construction of a fundamental logical contradiction within its own structure.

However, in spite of this apparently insurmountable logical objection to the extreme form of social constructivism we need to return to Smith's account in order to deal with a variety of other important issues which arise within the course of this general debate. Smith's first concern is with the way in which the representations of nature are, according to constructivists, in various ways reflective of 'the particularity of their origins in social circumstances' (Smith 2001: 115). This leads on to the claim that conceptions of nature are always contested. It is worth noting at this point, however, that these are two different points entirely. The concept of nature may be essentially contested, in the sense that how nature is conceived may be always a function of how it is fitted into an overall world-view, with no possibility available for the definitive resolution of which world-view is correct. It is a further, and controversial, claim that the construction of world-views is itself to be accounted for in terms of 'different histories, traditions, social practices, power relations and so forth' (ibid.). That may or may not be so. It has to be noted that the construction of such theoretical formations might rather be a purely intellectual endeavour on the part of individuals with a taste and aptitude for such activities which produce bodies of thought that may lie dormant until some social group or other takes them up as a means of justifying some practical, and probably self-interested, endeavour. It would be possible to view Smith's book in this light – and perhaps this one too!

The difficulty of evaluating large-scale synoptic views of some complex pheno-menon and determining which ones are the more reasonable to accept is what has given rise to modern philosophy of science (see Chalmers 1978). The struggle to elucidate the status of theories relative to observations, the role of experimental falsification, verification and empirical experiments, and so on has been conducted in a way which manages both to preserve the universal significance of scientific method and to allow the concept of scientific progress to retain some credibility. This suggests that essential contestability of concepts, arising from their embedding within complex synoptic views, need not itself cause us to abandon either of the main modernist aims of universal significance or a conception of progress. Hence the need to keep the concept of essential contestability (and its corollary of fallibilism) clearly separate from any doctrine of the social origins of such synoptic views, which may be no more than a sociological dogma of doubtful validity.

The conflation of these ideas leads on to a controversial conclusion. Smith himself draws attention to this conclusion, although not to the conflation on which it rests (Smith 2001: 116). One might be led to argue as follows. If the concept of nature is essentially contested then that is because the concept is embedded within a synoptic view. If all synoptic views are socially constructed, then all concepts of nature are socially constructed and to be explained in terms of the interests of social actors. Hence the value of nature, however construed, is always at bottom its value for some social actors or other. Hence, the conclusion is drawn, the concept of nature as having intrinsic value is not really intelligible. This then appears to block the attribution of intrinsic (that is, non-instrumental) value to nature. Smith notes the wrath this attracts from those he calls 'deep ecologists'.

However, if synoptic views are not to be explained along social constructivist lines, then this pattern of reasoning is itself blocked. This point leads on to the more general point that there is in any case a non sequitur in the reasoning of the last paragraph. For, even if all synoptic views are socially constructed, it does not follow that the concept of intrinsic value could not intelligibly form part of such a construction. It might be in the interest of some group responsible in some sense for the construction of the world-view that intrinsic value be attributed to some item or other. For example, it has been quite common for religious world-views to attribute intrinsic value to human beings, even though such views are, for all social constructivists, typical examples to take of socially constructed positions.

So far my comments on Smith's introduction of the social constructivists' position has been rather negative. The position, at least as put forward by Smith himself, is on my account open to various objections. There seems yet to be no very good reason to take it seriously, except as a cautionary point telling us to watch out for irrational prejudices within our own belief structures.

However, Smith then goes on to make the point that it is vital to distinguish between two different versions of constructivism (Smith 2001: 118):

1   as an epistemological thesis: our knowledge of nature is socially constructed;
2   as an ontological thesis: nature is nothing more than a social construct.

He then explains that, properly construed, social constructivists are to be understood as making the epistemological claim with regard to nature that our knowledge of it is 'inherently culturally bound', not the ontological claim that nature is 'nothing more than a cultural category' (ibid.: 120). Proponents of the latter he refers to now as 'strict constructivists' (ibid.: 121). However, even the latter are, on his view better construed as 'bracketing' issues about the world's ontology, than as making ontological claims. Since ontology is a function of the truth of statements about the world, this means in turn that strict constructivists are more concerned with understanding the social processes whereby ontological claims are made than they are with the truth or falsity of those claims.

However, this is to put the matter too mildly, for Smith tells us that they also hold that '[ontological] claims are . . . culturally bound and thus ultimately undecidable' (Smith 2001: 120). It is not clear what is being said here, for what cultural binding involves is not explained, but what does seem evident is that, on this view, ontological claims, since their truth or falsity appears to be beyond reach, are not rationally assessable, even from within the cultures that produce them. Hence all there is to investigate is how they are produced. An immediate rejoinder which is appropriate here is that it is hard to see how this can leave room for any fruitful discussion over environmental issues, which are chock-full of ontological claims and counterclaims, and thus it is hard to see why any environmentally concerned person should be interested in defending this form of constructivism. It is also unclear how a strict constructivist can give a truthful account of how claims are socially constructed, since a true account will give us ontological information – about the actual contents and operations of the social world (whether it contains social classes, individual consciousness and so on).

Of course, one claim which can clearly be extracted from these lines of argument is that different values dominate or prevail in different societies. In explaining this it is perfectly intelligible and plausible to suppose that it is in the interest of certain powerful groups within the societies in question that certain values should prevail. But this does not show that such views are 'socially constructed' in any interesting sense, rather than being socially propagated or suppressed. It does not in itself license a relativistic view of truth or meaning.

What strict social constructivism fails to allow for is any room for reflective self-criticism, for if all values are simply social constructs then it is hard to see where the necessary conceptual space is to be found for standing outside of them and criticizing them. But the human mind can range beyond what is given, can bring points of view into meaningful and creative conflict, whether or not the stimulus to this comes from cultures other than the one inhabited by the thinker. Orwell's Newspeak – a socially created language in which only approved-of thoughts can be entertained – is not possible (Orwell 1971: appendix). For a language within which any thoughts (as opposed to blind parrotings) can be expressed necessarily contains methods of self-reflection, negation and the expression of contrary-to-fact hypothesis. This is not to say that, within some cultures all of the time, and within all cultures some of the time, attempts will not be made to suppress undesired opinions, to inculcate beliefs, to influence what is to be regarded as desirable and undesirable. But that is a wholly different business, and one which is fraught with all kinds of difficulty, as the parents of teenagers can amply testify, from what is intended by the hypothesis of strict social constructivism.

However, we now arrive at what is really the heart of the matter, which comprises the positive reasons for urging deep ecologists to take on board the social constructivists' position.

Smith's first claim here is that constructivism has always played a vital part in the radical critique of oppression (Smith 2001: 125). He tells us that:

> Those in power always justify oppression, inequality and injustice as the necessary outcomes of universal, objective and unchangeable natural laws. Once this supposedly natural order is revealed as a social construct that is often arbitrary, always the result of special circumstances and everywhere serving the particular interests of the ruling elite, the possibility of change arises.
>
> (Smith 2001: 125)

However, this claim would need rather more careful examination and support than Smith gives it. Some social constructivists, such as Marxists, have themselves created vicious tyrannies, in part insulating themselves from the possibility of self-criticism precisely by characterizing their opponents as subject to the value- and belief-systems created in the interests of the dominant capitalist class. Marxist–Leninists in the USSR and elsewhere have perpetrated the largest intellectual and political disaster of the modern period, precisely by refusing to examine the truth in the claims of their pro-capitalist opponents until it was too late, and instead engaging in the practice of diagnosing them. If Marxism is ever going to regain its credibility it will have to jettison the version of constructivism found in Marx. Marxists will have to realize that, just because your opponents have a vested personal and ideological interest in criticizing you, it does not follow that their criticisms are not correct. The most effective criticisms you face will often be those of your enemies, in whose interests it will be to find the truth about the fundamental weaknesses of your position, because they often will be precisely the elements which you will find it hardest to hide.

In any case, if contribution to liberation from oppression is to count as a reason for deep ecologists to accept social constructivism, then it will have to be recognized that in practice many upholders of universalist or objectivist positions have also played a key role in such activities. For example, many religious believers have been effective in criticizing oppression within their own and others' societies – the Christian socialist, anti-slavery and anti-apartheid movements are a case in point.

We have to distinguish between two different possible kinds of critique of oppression carried out in the name of what purport to be 'unchangeable natural laws'. One is certainly the constructivist one – that there are no unchangeable natural laws, at least in the realm of human society, so a fortiori these cannot be such laws. But the other is to argue that it is the particular examples put forward by the oppressors which are not specimens of such laws. This, of course, allows that there may nevertheless be such laws. Short of detailed, and probably inconclusive, historical investigation, it would be difficult to tell just how many of the effective critics of social oppression are in each camp. In any case, it is also a further issue of how far their effectiveness, in terms of the actual ending of oppression, derives from their arguments, of whichever kind they are.

The thesis of social constructivism is at its most plausible when it is considering systems of values and their associated norms and prescriptions. It is least plausible when considering natural science. This is because natural science has developed a body of thought which is largely autonomous of society, in the sense that its truth or falsity is to be determined by agreed, society-independent procedures, designated as the 'scientific method', and the truth or falsity of the theories themselves are established independent of the cultural values of any specific social order. The fact that individual scientists are human beings with their own prejudices, values and self-serving propensities means that we need to distinguish between what scientists claim to be the truth and what science has shown to be true. It is the aim of the scientific method, among other things, to establish the distinction between these two.

These points about science are not meant to rebut in any way the perfectly valid criticisms made by many environmental critics of the use which has been made of science-based technology in the human alteration of ecological processes, usually for the worse. For example, Vandana Shiva has provided a telling critique of the misapplication of scientific thinking in the course of the 'Green Revolution' in India in the 1970s and 1980s (Shiva 1991). Her analysis of the failure of science-based crop development to take account of the adverse wider ecological effects of the introduction of high-yield varieties of rice and wheat in the Punjab, as well as of the equally adverse social and cultural ramifications of the change from traditional to high-intensity agriculture, is an object lesson in how a little learning can be a dangerous thing.

Such a danger is particularly liable to materialize when harnessed to powerful economic and political forces which are inclined to sweep aside objections to new technology as simple Luddism. It also has to be accepted that in practice the reliance of institutions of scientific research on lavish amounts of funding often leads scientists to justify their research in terms of the emergence of technology, which they are strongly inclined to trumpet either as of great benefit to humanity or as productive of lucrative market (and so of employment) opportunities – or, for preference, both. In such a context it is clearly tempting for the scientists concerned to ignore or downplay the possible adverse effects of the proposed technology.

Still, none of these crucial points about the sociology of science, and its location within the power structures of a competitive state and market system, needs to lead us to the conclusion that there is nothing to scientific theorizing which merits the designation 'knowledge', and important knowledge at that. Shiva castigates the science behind the 'Green Revolution' in various ways – as 'reductionist', falsely 'value-free' and decontextualized in its assessment of the impacts of the technology it produces (that is, it assumes that laboratory results will simply be replicated in the real world) (Shiva 1991: 21–2). In connection with the latter point she tellingly notes that the fragmentation of science into separate, hermetically sealed-off specialisms can militate against a proper scientific assessment of the complex impacts of technologies, which will often require the contributions of various disciplines.

But some of these charges are hard to sustain in the full generality in which they are delivered. Science may be 'reductionist' in some ways, but it is a matter of controversy among scientists whether or not there are emergent properties which cannot be reduced without remainder to properties of phenomena at more fundamental levels. Again, the issue of the extent to which factual claims can be distinguished from value claims is also a matter of debate within science. It is much more fraught as an issue within the social sciences, where understanding and explaining the object of investigation, human beings, requires an understanding of the value-judgements made by those objects, value-judgements with which the explaining theorist may or may not agree.

There is no good reason, however, for supposing that science, qua science, has difficulties with 'relational properties and relational impacts' (Shiva 1991: 21). Much scientific theorizing seeks to grapple precisely with these. After all, the paradigm of scientific theory and prediction, namely the application of Newton's laws of gravity to the movements of the bodies which make up the solar system, is highly relational and consists precisely of trying to specify enormously complex relational impacts. Insofar as scientists may oversimplify their understanding of the range of relationships to consider in a given instance as a result of the ignorance which results from their specialization in a subdiscipline, then the remedy for these defects seems to be better science, rather than a general suspicion of scientific theorizing in general, and the elevation of non-scientific understanding of the natural world to a comparable level of prestige.

After all, Shiva's devastating critique of the effects of high-yield varieties on biodiversity, disease resistance, soil structure and water sources (to say nothing of her social analysis of its interaction with the complex cultural situation in the Punjab) rests itself on a recognizably scientific analysis of the relevant data and a scientific explanation of the virtues of the traditional agricultural practices so cavalierly thrown aside. Traditional practical knowledge may not derive from the mode of investigation which has come to characterize modern science, but science is continuous with everyday empirical interaction with the natural world, and insofar as traditional practices do rest on knowledge not yet revealed by scientific investigation it should not be assumed that it rests on knowledge which the scientific method is incapable of unearthing and validating by its own means. None of which means that we should not, qua scientists, treat with great respect practices which have worked well even if (as is true of much successful technology) they have not been derived from the laboratory.

To return to social constructivism, we next need to acknowledge that there certainly are phenomena which are purely social constructs, in the sense that they exist only because of certain (often implicit rather than consciously undertaken) agreements between human beings living in a specific society, and they contain a strong element of the arbitrary. The phenomena in question are what is known as social conventions. These, as David Lewis has shown, involve the arbitrary selection between at least two alternatives of equally efficacious means to the solution of a specific kind of problem, known as a coordination problem (Lewis 1969). The easiest example of this to take is the problem of ensuring that accidents are kept to a minimum upon a highway with fast-moving vehicles by ensuring that they all drive on the same side of the road. There are two solutions to this problem – drive on the right or drive on the left. The choice between these is arbitrary, in that it does not matter which is chosen as long as all make the same choice. There will then ensue coordination of all drivers' behaviour in a manner which solves the problem.

The choice is thus essentially arbitrary, although at least one of the solutions is likely to be 'salient' for a given society because they have some reason, independent of its effectiveness, for finding it more 'obvious' than its rivals (perhaps it calls to mind an earlier solution that is 'mutually known' by all its members). However, although the choice between the (at least two) solutions to the coordination problem is arbitrary, and thus the convention which emerges contains a central element of arbitrariness, this arbitrariness occupies a very specific place within the phenomenon of the convention and does not mean that that whole phenomenon is arbitrary. We can see this when we return to consider the driving convention example. What is obvious is that it is the laws of physics which set up the coordination problem in the first place. It is a fact about the way the world is that solid objects which impact on each other at high speed usually inflict various amounts of damage on each other. It is not a matter of convention that this phenomenon exists, even though the emergence of a convention is an excellent solution to its effects in the case of fast-moving solid objects containing human beings.

It follows from this analysis that a convention can only exist where there is a coordination problem with at east two solutions between which choice is arbitrary – one is as good as the other, so tossing a coin to decide which to choose would be a perfectly rational way to make the choice. But where a solution is chosen because it is uniquely the best solution, then we do not have a convention, even if what the problem is and what the solution is are both unique to a particular society. If it turned out that, for some hitherto undiscovered reason to do with human physiology, drivers made fewer accident-causing errors of judgement when they drove on the right rather than on the left, then there would be good reason for all societies to adopt the former rule. The choice would no longer be arbitrary, and any society which chose the other alternative could intelligibly and probably fairly be accused of endorsing a foolhardy practice.

Conventions, then, are very specific social phenomena. They are clearly very important in human life. For example, as Bennett has ingeniously argued, the coming into existence of human languages may be analysed with the use of Lewis's theory as arising from the arbitrary selection of certain sounds for the purpose of communicating speakers' meanings (Bennett 1976). But this does not imply that the truth of what is said using such conventions is itself a matter of convention. Clearly, if Bennett and other similar theorists are right, the way the world is can be grasped only through the deployment of some conventions, such as the aural and visual signs arbitrarily selected to do the business of representing the world. But the 'way the world is' is not itself a matter of convention. One way, then, of understanding social constructivism is to see it as incorrectly extending the notion of a social convention beyond its proper bounds.

However, it might well be replied to these points that if they work they do so by focusing on social constructivism as applied to factual claims about the world. Do they apply with equal force to value-judgements? And if it should turn out to be impossible to differentiate between fact and value, may not the social constructivist argument overall still be vindicated?

We can be objectivist about matters of fact without any immediate show of implausibility – it takes some rather sophisticated philosophical analysis to produce the basic case for social constructivist positions here. But it is readily apparent that objectivism about value-judgements looks odd almost from the start. What critics of social constructivism will need in this sphere is some form of intersubjective validity for value-judgements. An at least partial grounding of this on a naturalist position might give such critics much of what they want. Human naturalism would be the model here, in which value-judgements in the area of ethics are argued to converge among all people when they adopt the appropriate social standpoint, set aside personal interest, and exercise their common capacity for sympathy. Then their shared human nature will be found to produce a commonality of emotional response which is the basis of judgements of virtue and vice. This also gets us from 'is' to 'ought' without the intervention of reason, but in a way which makes intersubjective agreement in moral judgements possible (Selby-Bigge 1902: 169–294).

Some environmental ethicist critics of social constructivism, such as Rolston, have attempted instead to locate value objectively in nature by demonstrating that (non-ethical) valuing occurs in species other than the human (Rolston 1997: 59–63). This is correct, but does not show why human beings should agree to accord any ethical value to this valuing. What he would need to defend his position is some basis for claiming that human beings will converge in their judgements concerning the value of nature. Perhaps once they have adopted the appropriate standpoint, and their common nature, involving perhaps biophilia (the evolution-arily produced responsiveness to nature which E. O. Wilson has postulated (Wilson 1984)), has come into play, this convergence may tend to emerge.

Such an approach suggests, too, the area where Smith and I differ over our understanding of the nature of ethical thought and argument. Smith asserts that 'Ethics should be bottom-up, beginning with the phenomenology of actual ethical feeling.' He sees a role for social constructivism in the attempt to 'understand and express the "place" of ethical values in relation to the social and natural circumstances of their production, to interpret their meanings, their emotional resonance and so on.' He also regards society 'as a constitutive part of the *production* of ethical values, not just the filter that facilitates or delimits the recognition of values that are, in some sense, already there in nature.'[2]

It is certainly acceptable to grant to ethical feeling a fundamental place in ethical thought and argument, and to accept that specific value-systems are produced largely in ways which give social forces, such as socialization, an important role. We should also accept that in actual debates about values it is important to start from the value positions to which individuals are actually committed, and to seek to move ethical debate on from that point.[3] But we need also to affirm, as Smith himself does, that for human individuals epiphany events can occur in which value-judgements are made as the result of the direct, non-socially mediated response to something (Smith 2001: 167). Such experiences may lead to the adoption of ethical positions at odds with those prevailing within the society within which the individual has reached maturity.

A universalist, thus, ought to consider seriously the advantages of a naturalistic account with respect to the origins of ethical feelings, which countenances a universal structure of ethical feeling common to all members of the human species, a structure which forms the basis for interpersonal agreement on matters of ethical value. However, as Hume noted, actual ethical feelings can be distorted by a variety of factors. His favourite example is that of partiality – which may, for example, lead us to fail to have the feeling of disapprobation which is the natural response to an anti-social act when it is our own loved ones who are the perpetrators (Selby-Bigge 1888: 582–3). Hence, actual ethical feelings are only ever data which cannot be taken at face value. They can be appropriately criticized by means of ethical reasons which place them in a different light. Of these reasons, the ones which are designed to get the owner of the feeling to view the matter judged in a universal, impersonal light, are the most characteristic and time-honoured.

The facts to which Rolston alerts us in his argument concerning the valuing activities of non-human organisms may well have this role, in getting someone who is inclined to view such organisms as possessing no ethical standing to note ethically significant similarities between the organisms in question and human beings, ones which they may have failed to notice, or perhaps not have taken full account of (indeed, this kind of point is one upon which we will be setting great store in the development of a theory of ecological justice). This may be designed to release the appropriate ethical feeling, so that the judgement of ethical worth is issued without further reflection. But it may also be that this move does no more than cause the interlocutor to recognize an inconsistency in their position which they resolve by adjusting their judgements accordingly. This might itself be deemed to be a matter ultimately of feeling rather than of reason, for unless one cares about inconsistency one will not be led to make the change. However, this is arguably not in itself an ethical feeling, but rather another natural kind of feeling (which may well have a naturalistic explanation) of which use can be made in ethical debate.

We should, then, if we are universalists, be sympathetic to the view that human beings' ethical judgements rest on a naturalistic basis of feeling. In this we can properly share Rolston's concern to prevent the field of value from dissolving into endlessly shifting social constructions which are conceived to result wholly from factors entirely internal to social life, even though the particular way in which Rolston chooses to do this has to be rejected. We should be willing to accord social forces and pressures – some systemic and impersonal, some the result of deliberate manipulation – an important role in bringing about the specific ethical judgements which characterize actual human beings in concrete situations. But there is good reason to posit a naturalistic basis for interpersonal agreement once the work of reasoned argument and social critique has been engaged in. This stems from shared feelings both about matters of ethical value, and also about more intellectual matters, such as consistency in judgement, for both of which it at least makes sense to try to find a naturalistic and ultimately, it can with increasing plausibility be argued, an evolutionarily-based explanation (see, for example, Pinker 2002).

However, at this juncture we should return to Smith's specific account of the problems with universalism in moral thought in order to seek to vindicate this kind of theory against his contextualist approach.

3 Contextualist rather than universalist and rationalist morality?

We must now examine the positive characterization of moral thought and its relation to the natural world which Smith develops in the remainder of his book. He favours an approach which he dubs 'radical environmentalism'. This is fundamentally at odds with the modernist world-view, which embodies the post-Enlightenment commitment to a neutral, universalist model of reason understood as employed in natural science, analytical philosophy, the legal and bureaucratic structures of the state and capitalist market economics. It is a view of reason that ignores the historical and culture-bound bases of thought which, as we have seen, Smith believes it is the strength of social constructivists to have focused on and illuminated.

By contrast with this position, characterized as antithetical to 'radical environmentalism', we can trace the positive characterization of the latter which emerges in the book. We can summarize this characterization as follows:

1   Radical environmentalism avoids entanglement with scientific modes of thought and the falsely conceived model of neutral reason on which it is said to rest. It thereby avoids the pitfalls of a naïve naturalism, which is reductivist and regularly entwined in reactionary politics.

2   It avoids the model of moral thought as the rational exposition and application of falsely conceived universal principles to particular situations. Instead, it focuses upon the 'non-rational' elements in human moral reactions, such as love of a particular place, and emphasizes the importance of responding to the objects of moral concern in all their context-bound particularity.

3   It rejects the idea of authority in all its forms – moral, legal, political, religious. These sources of 'law' are all regarded as coercive and harmful impositions of the interests of the dominant members of hierarchically organized groups upon both human and non-human victims. Thus, it espouses antinomianism (anti-law) and, in political terms, favours direct participation in anarchistic structures.

We have seen in the previous chapter the way in which Smith argues for (1). He argues for point (3) in chapter 5 of his book. We need not spend too much time examining those arguments, the discussion of which would take us too far from the direct concerns of this book. They are largely a matter of noting the parallels between present radical environmentalists and antinomians of the past, such as the Levellers and Diggers of the English Civil War. We should, however, note the characterization Smith gives of the antinomianism favoured, on his account, by radical environmentalists:

> From this historical sketch we can see that the key constituents of antinomianism might include: the *rejection of political authority*, in particular as it is embodied in the law and associated institutions; the *rejection of all moral authority* and a consequent refusal to accept the *imposition* of moral norms whether secular or religious; an emphasis on the *free individual* as responsible for creating their own space of engagement with the world; a deep-seated suspicion or even outright rejection of *rational authority* insofar as it claims to be the sole arbiter and administrator of our lives; and last, but not least, a utopian belief in a *New Jerusalem*, a new world *without an order* and without hierarchies.
> (Smith 2001: 138; original emphasis)

Let us, however, focus upon point (2) above, which encompasses the 'ethics of place'.

Smith begins chapter 6 of his book with the social constructivist claim that it is at the meta-ethical level as well as the first-order level that cultural structuring takes place. As he says, 'Theories of value are never, whatever they may claim, entirely abstract or context free' (Smith 2001: 151). He then sets out, using a 'spatio-temporal metaphorics' (ibid.), to characterize the differences between the modernist and radical environmentalist meta-ethics.

There then follows a characterization of modernity in terms of a whole series of social institutions and practices prevalent in the contemporary world. The beginnings of Smith's critique of modernity appear when he seizes upon the concept of 'progress' and points to the continued failure of the modern world to produce the predicted improvements towards which its ceaseless change is supposed to be oriented. The result is a redefinition of 'progress' as 'what change in the modern world produces'. Hence, the chimera of progress is absorbed into an ideology of change and the instrumental rationality which is applied to create change. The ideal future is constantly deferred, but the all-powerful engines of change whose existence is justified by that ideal so operate as to put the continued existence of the world itself in doubt (Smith 2001: 154). Out of this hyper-Heraclitean flux emerge the characteristically anomic, atomized individuals, fixated upon change and newness for its own sake and subject to consumerist fantasies, so characteristic of contemporary Western societies.

This is a diagnosis upon which many environmentalists, but not just them, would be happy to converge. However, recognition of its truth (if we are still allowed to use this term within social constructivist perspectives) requires that those to whom it is addressed, many of whom are the social products of these very societies, recognize that progress has not been achieved, that it is constantly deferred, and that change has replaced the always-deferred ideal as the de facto point of the whole modernist structure.

In the light of the general social constructivist project which is being espoused in these pages, many questions become pressing. For example, how is this recognition to be achieved except via the application of a critical mode of thought which compares projected benefits with actually achieved non-benefits, notes the disparity, and then looks for an explanation of the failure to achieve the goals set? To do this from within a cultural structure which is said to contain powerful forces influencing people in the self-justificatory direction appears to require some ability to achieve distance from that structure and to find resources to criticize it. How is that possible? More specifically, if all perspectives are culturally produced, what cultural elements are producing that critical perspective? Do we have to posit culturally alien elements within modernistic societies which subvert it from within? Or is it the encounter with alien systems, that is, with those occupying a place outside the culture which is criticized, which is the source of the critique? Or is it some internal contradiction within the culture of modernity which throws up its own self-critical moment? Specifically, where does Smith get these ideas from?

Where, also, does the doomsday prophecy with which Smith concludes his diagnosis of modernity come from – the claim of environmentalists that 'a *world without ends* . . . is . . . likely to entail *the end of the world*' (Smith 2001: 154)? Insofar as this is meant to be an anticipation of the breakdown of the biospherical processes upon which all life is said to depend as a result of human beings' rapacious approach to the natural world, this looks suspiciously like a science-based critique. In fact, for most environmentalist critics of modernity that is precisely what it is.

Smith next counterposes an example of the 'ethics of place', namely the system of 'tradition and local agreement' which governed the mediaeval 'commons' (Smith 2001: 155), with the two dominant modernist ethical systems – utilitarian and rights-based. These have been crystallized into social practices of impersonal bureaucratic rationality, as described by Weber. Those who adopt the adjudicative bureaucratic roles within this system are meant to apply impersonal, abstract, universal, context-free principles and rules to arbitrate moral rights and wrongs. The upshot is a socio-ethical system which is 'concerned with interpellating the individual into an already given social framework and promulgating a top-down set of values to be internalized rather than letting values form and operate from the bottom up by communal participation' (ibid.: 157). Smith also makes the standard connection between Benthamite utilitarianism and neo-classical economics, both of which reduce problems of social choice and welfare to an individualized calculus for determining the distribution of a homogeneous element – pleasure or preference-satisfaction.

Rights theories also 'bear witness' (Smith 2001: 158) to the privatization of the ethical – that is, its reduction to a series of individual claims to bundles of rights and happiness. Smith's spatial metaphorics draws the comparison between the enclosure of the commons and the individualization of the ethical, although both are seen as having the same underlying connection with modernism.

Throughout this account, as is indeed the case throughout the book, the social constructivist analysis of the cultural formation of value- and belief-systems involves the employment of phrases which are highly slippery. They insinuate, but do not overtly provide, explanations of the connections between ideas and social institutions. For example, we are told that 'We have a tame ethics, an ethics of control that *encloses* the moral field via a process exactly analogous and closely allied to the enclosure of Britain's social and agricultural spaces' (Smith 2001: 159, italics in original). But we are not told why weight should be placed on this analogy, or in what the 'alliance' consists. Earlier we are told that 'The parcelling up and privatization of land goes hand in glove with the reconstitution of the ethical sphere along similar lines' (ibid.: 156). The metaphor suggests, but does not explain, an intimate, perhaps causal, connection between the 'hand' and the 'glove'.

More generally, we are told that there are 'links' between moral values and 'forms of life' (Smith 2001: 151), that certain ideas are 'products' of particular times and places (ibid.). These expressions and others like them (such as 'corresponds to', 'related to', 'mirroring', 'embodying', 'echoing and supporting', 'have much in common') always emerge when social constructivist accounts are proffered. They are not explained, except in terms of the general claim that ideas are cultural products. Hence, we have an enclosed mode of thought, in which apparent explanations are always given in terms of overt or covert metaphors, which are in turn held together by their common relation to the generalized and very vague claim which expresses the social constructivist thesis.

All the targets at which Smith aims his critique are certainly justifiable ones from the environmental point of view. For example, utilitarianism systematically restricts the scope of ethical concern to sentient beings or, in its neo-classical economics variant, to beings capable of expressing preferences in the market-place. These are open to serious objection. But to show this we have to fight them with the appropriate weapons, which are elements of ethical theory. We have to show, for example, that the category of the morally considerable extends beyond the purely sentient, to encompass all beings possessed of the capacity for well-being, which on some views encompasses ecosystems and species too (Johnson 1993: ch. 4).

Of course, this requires that we employ the techniques of philosophical argument and the full panoply of logical critique, that we search for inconsistencies, and so on, which Smith has already impugned as being 'bound up' with the harmful modernistic rationalism of the contemporary world. It is also true to say that such arguments, being philosophical, will seldom be conclusive. But, given the rather unsatisfactory state of the social sciences, it can scarcely be supposed that sociological arguments are going to be any better, and in their social constructivist form it is hard to see that they really amount to arguments, rather than insinuations, in any case.

It is also certain that the anarchistic and antinomian position which Smith seeks to vindicate will need to address some serious philosophical and practical objections, which cannot be properly met by the riposte that those who put them forward are expressing views which 'correspond' to dominant social practices. For example, how is a social system of any complexity, containing the billions of human beings who currently exist within mass societies, in which the over-whelming majority are strangers to each other, to operate in a manner in which individuals are to be free from domination, yet jointly determine their social practices by procedures for participatory democracy? You do not have to be a doctrinaire capitalist individualist, or the dupe of such, to raise such a question.

Let us now return to Smith's positive characterization of the 'ethos of radical environmental action' (Smith 2001: 160). The ethics of place holds open the possibility of giving 'voice to concerns that are otherwise marginalized or mis-represented in hegemonic discourses' (ibid.).

The example of various forms of radical environmentalists' 'direct action' is given as a way of contrasting their non-instrumental, contextualized, unstructured modes of thought and action, involved in, say, finding various ways of delaying road construction, with the utilitarian, instrumental, rights-based, one-dimensional and context-free mode of thought and action of the roadbuilders and their bureaucratic, coercive, regimented bureaucratic guardians of police and planners.

Modern(ist) road-systems themselves are contrasted with the highways of earlier periods. The former are one-dimensional constructs designed solely for speedy movement from place to place. They are highly regulated, and subordinate every consideration to the needs of the motorized transport which they bear. The latter were multi-dimensional, localized public places in which it was possible to meet others, stop and talk or engage in other social interactions. Their forms and direction would be the result of historical happenstance emerging from the particular locale and context in which they were situated. Modern roads are mythologized as spaces of freedom and individual choice. The reality is that they do not even deliver on their own promises, as traffic snarl-ups become everyday experiences and the impetus to deliver on the undeliverable promises leads to ever-more environmentally destructive bypasses and new roads which also quickly become congested (Smith 2001: 160–2).

In the course of this eloquent and persuasive diagnosis, Smith remarks that 'We find that we are forced to restructure our lives to fit the needs of what was supposed to serve our interests' (Smith 2001: 161). Hence, he claims, we should note that devices supposedly designed as neutral means to achieve whatever ends their users see fit to pursue are always in fact manifestations of 'partisan politics' (ibid.). The road requires cars (and other such machines) to be used on them. Foot transport, horseback, bicycles and other alternatives are either extremely dangerous, inconvenient or expressly forbidden. The favoured modes are then celebrated as expressions of national greatness and national progress. The alternatives are marginalized as quaint, old-fashioned, eccentric or retrograde. There is no genuine neutrality between people's choices, but a subtle or not-so-subtle dragooning of them in economically desired directions.

However, this analysis, though telling, also might properly be construed as itself partial. This, of course, may not matter to Smith's project, which is to find a 'space' within which to articulate alternative ethical visions to those which he has located firmly within the modernist structures of contemporary societies. Nevertheless, his analysis is in danger of overlooking some of the complications of the relations between human beings and their technology.

An artefact, such as a road or a car, necessarily is independent to a certain degree from the human beings who create it. It needs to have this independence in order to be of use. An artefact which has primarily instrumental value for the user, a tool in short, is designed to save some kinds of effort, to economize on time, space, energy or some other valued but relatively scarce item. Hence it is made to do its work, at least for a time, independently of the human beings whom it is designed to serve. A road saves human users from having to expend time and energy cutting a way through natural obstacles to movement, for example. Hence it needs a certain degree of permanence, to stay in existence ready for use after its initial creation. But this, of course, comes at a price. It will require new kinds of human effort to maintain its proper working, thereby securing the advantage for which it was created. Another aspect of the price which its creation exacts is that certain opportunity costs are imposed. The space which it occupies may not be usable for other desirable purposes, and the time and energy devoted to its maintenance will inevitably take away certain other possibilities to which those resources might otherwise have been devoted.

In both of these ways artefacts do indeed 'force us to restructure our lives to fit the needs of that which was supposed to serve our interests' (Smith 2001: 161). However, that is, for the reasons just given, inevitable. All tools must do this to some degree. If the costs in time and effort required to maintain a tool's existence and proper working are too great, and the opportunity costs which it imposes are too onerous, then we might lament its creation rather than celebrating it. But we should not suppose that we have the option of creating tools which do not in some way dominate us to some degree insofar as we make use of them.

Hence, the structures of modernity, which have harnessed science to technology to a very high degree, at best differ only in the extent to which they represent impositions upon their human creators as compared with their pre-modern precursors. We can escape this if we eschew tool use altogether, but given the paucity of our purely bodily resources that does seem a recipe for a life which is nasty, brutish and short.

However, in making these comments, which are intended to be uncontroversial, we have inevitably had to make use of an instrumental language of accounting, referring to prices, costs and benefits. This does seem genuinely to be unavoidable – instrumental reasoning is expressly designed to assess instruments. Human beings have employed this language without much hesitation throughout their history, and in all times, places and cultures have been willing to give up their existing tools for new ones which promise to produce a better outcome. Sometimes this has been done with regret for the passing of the old ways bound up with a given set of tools, sometimes in the face of objections from others of their contemporaries based on religious or economic concerns, as with the Luddites and the Amish.

This thought is at the heart of any materialist conception of history, and is central to the conception of human activity as praxis, which conveys the idea that in remaking their world human beings remake themselves. We certainly then need to stand back as best we can and try to assess whether the world which we are making, the making of which requires a remaking of ourselves, is imposing too great a loss of opportunity costs, and too severe an alteration to our human character, for the degree of benefits which it is supposed to be providing. But in seeking to do this, which is what in large part Smith is trying to get us to do in his critique of modernity, we do not escape the necessity of adopting the accounting language which he wishes to brand as modernist and biased in favour of the environmentally destructive status quo.

An ethics of place which seeks to relate self and world in a richly contextualized way will itself impose costs. It will require a different praxis, a different set of tools (since the 'no-tool' scenario is too horrendous to contemplate) and a different application of instrumental reasoning. But it will still be embedded within the same overarching conceptual structure which has been found unavoidable in all cultures. Hence, the ethics of place will give its own account(ing) of the costs and benefits of a richly contextualized mode of praxis, either overtly or implicitly.

Nor is this to be construed as a political ploy or tactic, as Smith suggests is the case when radical environmentalists use the modernist rhetoric of 'rights' to express some of their ethical claims, and sometimes use the law to defend interests which they believe modernistic culture systematically ignores. He cautions us against taking these tactical ploys for an endorsement of the bureaucratic/legal structures within which the concepts of 'law' and 'rights' reside.

However, it would be a mistake to suppose that this interpretation can be made to apply to the accountants' language of costs and benefits, even within the ethos of radical environmentalism. Even if the language of 'rights' and 'costs' and 'benefits' is newly minted with the onset of modernity, these concepts have their non-modernist exemplars and precursors. Wherever people think that they have a justified claim upon the consideration of others, wherever they have demanded certain forms of treatment of themselves and their loved ones, wherever they have tried to work out whether the game is worth the candle, they will be employing these concepts.

If this is true, then the attempt to steer clear of such instrumentalist reasoning and concepts on the misguided grounds that they are uniquely modernist conceptions may account for the feature of radical environmentalists' expressions which Smith notes, namely that they are often vague and indeterminate. Smith puts this down to the fact that the radical environmentalists are using a language which 'makes certain things difficult to say (and apparently even more difficult to hear)' (Smith 2001: 164). This is the 'newspeak' hypothesis which we have already seen reason to reject. Smith's own text is an eloquent refutation of the whole idea. It is perfectly possible to express the thought that the modern world has involved the creation of a technological imperium which fails to provide the benefits it promises, and in the process destroys both the natural world and much of value in human life, desensitizing human beings and destroying the basis for a richly communal, non-coercive life for them. The problem is not in expressing such thoughts – for whole reams of this kind of analysis have been written, not simply by environmentalists. The problem is getting people to take them seriously.

This highlights the difficulty with looking at the direct action of radical environmentalists as any kind of model of a new ethical space and mode of existence.

Their activities are almost entirely reactive and focused upon a specific goal – to prevent something from happening. What is needed is an exemplar of radical environmentalist praxis in which the alternative instrumentalities and associated modes of living are on display. These need to reveal an alternative set of social (sexual, child-rearing, educational, spiritual), economic (tool-making) and political (decision-making) activities which embody the individualized contextualism of both life-styles and ethical thought to which the proponents of environmental radicalism give their allegiance.

Smith does go on to address this question, finding within the anarchic camps of radical environmental protesters and the 'mixed communities' of humans, bears, sheep and wolves in southern Norway, which Arne Naess has described, places where one may find the living examples which point the way to an alternative mode of existence to that of modernity (Smith 2001: 164–5). These are sketchily described, but the main point which emerges from their description is that they embody a way of relating to the world which eschews formal rules and codified ethical systems, and replaces these with a new attitude, of caring and inclusion, with respect both to human beings and, crucially, to the non-human world. The impression one has is of each individual within such communities searching for his or her own solution to such phenomena as clashes of interest (such as those between farmer and bear with respect to the sheep) on the basis of whatever considerations appear to that individual to be relevant to the particular case in hand.

However, lest this appear to be a matter of arbitrary whim, which looks like a recipe for despotism, Smith cites approvingly Naess's idea that there must be, within a biocentric society of the kind which he favours, a clear statement (presumably by individuals) of values, rules and norms. However, he interprets these as a 'loose framework', not a 'codified legal/bureaucratic system' (Smith 2001: 166). What guides one's application of these rules in particular cases is not a system of casuistry, but an attitude of caring towards others, including other species.

What strikes one again at this point is that what has just been described is really no different from the way in which traditional moral theories have been applied. Rules are never assumed to be straightforwardly applicable to specific situations, even in utilitarianism. There is always an irreducible element of judgement, covered in 'prima facie' and '*ceteris paribus*' clauses. In any case, there are modes of ethical reasoning – virtue ethics, situation ethics, for example – which are also present within 'modernist' societies and which look rather more like the ethical mode of radical environmentalism than utilitarianism. Smith appears to be oversimplifying his account of the ethical thought of 'modernity' to make the desired contrast.

However, his account of how one might be brought to adopt an attitude of caring for the non-human world is valuable and convincing, not least because he here breaks away from the social constructivist position. He characterizes the 'gestalt' switch to a new way of seeing our place in the world as often brought about by epiphany events, such as witnessing the destruction of a wood in the name of 'progress' (Smith 2001: 167).

Yet even here there is a worrying confusion of concepts. Smith notes that the 'depth of feeling' for the environment which is often revealed in such epiphany moments cannot be expressed in 'discourses of formal rationality and rights' (Smith 2001: 168). This may well be true, but that in itself does not show anything about the importance of such discourses. Their job is to articulate specific moral claims with respect to something once its moral standing has been recognized. Smith is undoubtedly right to extol the vital importance of the epiphany events he cites in getting that status recognized. But he may be wrong in supposing that that is the only way for the gestalt switch to be brought about. The deployment of rational argument which reveals inconsistencies within the treatment of different creatures may also be helpful for some people, and in any case will have some part to play even within context-laden moral thought of the kind which he praises – unless there are to be no serious moral dilemmas in the radical environmental view of the world!

There is another possible significance of the events which Smith mentions. In certain respects they are rather like falling in love (in fact, they may actually be examples of falling in love). Falling in love is the archetypal instance of the romantic attachment to another individual qua individual. One can see how an appeal to such feelings leaves itself open to the kinds of excoriation we may apply to claims of love between people (it's not love, it's juvenile infatuation; you're in love with the idea of being in love; you're confusing love with feeling sorry for someone; you've a romanticized idea of what she's like; the reality is very different, and so on).

But supposing the love is open-eyed, mature and not self-deceived, then it comes into a category which really takes us beyond ethics entirely. For then the object of one's love is experienced as having a claim upon one which cannot be adequately expressed in terms of obligations, duties and rights, even though these will usually be attached to the relations between loved ones. As Hegel noted, when marriage partners start talking the language of rights to each other the marriage is over (Hegel 1952: 110–11). As Bernard Williams also has noted, the relation one has to someone one loves will lead to forms of self-sacrifice and also to the giving of priority to the object of one's love which cannot be explained or justified in terms of duties and rights (Williams 1981: 18).

Conceivably this is what is being referred to in Smith's description of a gestalt switch – a non-rational 'locking-on' to an object which holds one's affection with great strength. This is indeed a powerful force in human life. But it is not something which can easily find its focus in the natural world as such. When it occurs it is necessarily a particular natural object which is its focus, such as a specific tree, wood, stretch of coast and so on. Once that 'locking on' has occurred, however, then one has a basis for expanding the ethical vision of the individual to encompass all of nature. But in doing this the deployment of rational argument seems unavoidable. Sometimes people deploy rational argument for themselves and reach their ethical conclusion unaided. Sometimes the conclusion emerges in the course of discussion with others. But the ethical generalization which is embedded in such universal ethical standpoints as Naess's version of deep ecology requires an intellectual grasp of the object of moral concern which cannot be accounted for simply by the epiphany experience of one element of it. Reason and emotion work in tandem here. Smith does a valuable service in emphasizing the non-rational element and its crucial role in the gestalt switch. But his anti-rationalism, deriving from his hostility to modernism, seems here to be misplaced.

At a later point in the argument Smith elucidates the alternative to a reason-based view of ethics. Specifically, he refers to Irigaray and her concept of 'wonder' which is said to be 'the elemental passion that can compose a space for difference, the "thin air" that separates us and allows us to keep a respectful distance' (Smith 2001: 183). The key point about wonder is that it involves a recognition of the way that the object of wonder exceeds/escapes our capacity for rational categorization. It is a tribute to the irreducible individuality, and thus difference from others, of the object. Thus, we are told, 'Ethics is then the flow of *things* in desire and wonder, it is a relation that lets things *be*, conserving and sustaining them in love and/of [*sic*] difference' (ibid.: 184). Once again, modernity's aim is regarded as being the squashing of this sense of wonder, as a sign of a long-banished immateriality. It is replaced with the homogenizing rationality of the general category and universal principle, common to both natural science and the rational bureaucratic order.

Here we are at last on ground which bids fair to illuminate the ethical response to the non-human that is the hallmark of radical environmental thought. I have argued elsewhere that the sense of wonder, and of wonderfulness, with respect to the non-human does indeed underpin our sense of its ethical dimension (Baxter 1999: 67–70). But in some ways the discussion of 'wonder' is itself a bit one-dimensional here. A sense of wonder certainly can be evoked by the experience of difference, uniqueness, individuality. But it can also be evoked by the sudden recognition of the nature of something with which one is completely familiar but has not properly examined before. One can feel a sense of wonder at formal qualities, such as complexity, or simplicity, or complexity-in-simplicity. Science and reason can themselves be objects of wonder. The sense of wonder has not disappeared from the modern world, but arguably the objects of wonder are no longer what they once were, because the objects themselves are no longer so easily available in a world of artefacts, and wonder has been increasingly redirected to human beings and their works to the exclusion of the non-human. That is, it is not a deficit of wonder from which we suffer, but a deficit of experiences of the objects which are the appropriate recipients of our sense of wonder, and of the ethical dimension to our responses. Certainly there is a good case for saying that the evincing of a sense of wonder towards the natural world is something which modernist ideology, with its determination to view the natural world as a bundle of resources, has done its best to block and ridicule. But that is not the same as saying that wonder has no part in modernist thought.

As noted above, it is arguable that what is being discussed here under the heading of 'wonder' might also, and perhaps better, be discussed under the rubric of 'love', which is also directed upon, and wholly valuing, of the unique and unassimilable otherness of its object, when properly conceived. But this points to one difficulty in the project of basing an ethics of care upon wonder/love for the unique 'other'. As we noted earlier, love arguably transcends ethics, rather than grounds it – or perhaps it transcends it as well as grounding it. Love makes exception, is self-sacrificing and in many ways is wholly amoral precisely because it works with the category of the 'wholly different/exceptional'. When love is the motive, justification is irrelevant, and justification lies at the heart of moral deliberation and debate. If wonder is like love in this respect, then it needs to be tempered, brought down to earth (a useful pun), to enable us to think our way through highly complex moral issues. The usefulness of this Irigaray-inspired focus upon the importance of the sense of wonder is, then, that it facilitates the escape from the prison of social constructivism. But it needs very careful treatment before it can furnish us with a defensible ethical position.

Smith notes Irigaray's comparative indifference to non-human nature as her examples of the 'other', but still sees value in her approach as avoiding the 'valuable because like us' and 'valuable because I can merge with it' views prevalent in defective forms of environmental ethics – the former characteristic of the ethical approaches of analytical philosophy, the latter characteristic of deep ecology. We need instead to care for the other, while respecting its unlikeness, even indifference or hostility, to ourselves. Smith ends this part of his discussion by endorsing the epiphany, and so unpredictable and non-rational, nature of the occasions when wonder opens us up to the otherness and ethical standing of the non-human other, and once again contrasts this phenomenon that the dessicated and petrifying categories of modernist, analytical ethics (Smith 2001: 188–9).

However, in spite of the advances made in this chapter beyond the problematic positions of social constructivism, what we end up with is an important, but, as noted above, not wholly adequate basis for developing an environmental ethics. In particular, we need a fuller explanation of how an ethics of caring, which rests on epiphany experiences of wonder and which respects the other in all its difference, can tell us how to balance the clashes of interests between those others. Until we have a clearer account than we have been able to glean from Smith's barely developed examples of mixed communities and radical environmentalists' camps, then we cannot say that we yet have a clear alternative to the universalist, modernist ethics which he so frequently excoriates.

Before moving on from this contextualist approach to morality it will be useful to take a brief look at one other context theory, in order to assess its treatment of an example of direct relevance to the topic of this book. It arises in chapter 6 of Karen Warren's book *Ecofeminist Philosophy* (Warren 2000). This chapter discusses the case for universal moral vegetarianism – another clear example of universalism – and presents her reasons, based on contextualism, for rejecting it – that is, for holding that there is no universally valid case for holding the eating of the flesh of animals to be morally wrong. The view which she supports – contextual moral vegetarianism – is analysed in terms of four propositions:

(1) Reasons for moral vegetarianism as a practice in a given circumstance will be affected by contexts of personal relations, gender, ethnicity, class, geographic location and culture. (2) Moral vegetarianism is not a universally

required practice in all contexts. (3) In principle, morally acceptable food-eating practices should not replicate or reinforce unjustified Up–Down systems of domination based on the power and privilege of Ups over Downs. (4) An ecologically informed care-sensitive ethics approach is helpful in unpacking the nature of our relations to non-human animals and to resolving contextual issues of moral vegetarianism.

(Warren 2000: 133)

There is much that Warren says in explanation and defence of these positions which there is not here space to examine in detail. But it is important to note that, at least in form, each of (1) to (4) has the appearance of a universally valid proposition, that is, they are not themselves contextualized. Further, the universalism which Warren appears to be attacking is arguably rather a straw man. Moral prescriptions can perfectly well have the form of universal claim and yet be subject, as noted earlier, to ceteribus paribus clauses expressly designed to recognize that in a specific context there will often be various competing moral claims in conflict, with the conflict having to be resolved by the rejection of some in favour of others. This appears, for example, to apply to those situations of necessity of which Warren cites as one example illustrating claim (1) – the vegetarian mother who kills an animal to feed her starving child. This does not contradict the universal validity (if it can be established) of moral vegetarianism as a general practice. Other examples which Warren gives, such as the cultures which view human beings and animals as part of a single ecosystem such that humans are food just as much as are other creatures, and thus which see no problem with eating animals for food, are examples of competing visions of the world. Their devotees undoubtedly see this way of viewing things as possessing universal validity, so that we are back with the standard problem of the different moral implications of competing world views. This does not show that universalism in moral thought is false; rather it shows the difficulty of deciding which universally claimed view is correct. The concept of a 'cultural context' is of no help here, since we know already of the view in question that it is a 'deeply held view' among members of certain cultures. But, for reasons which we have already examined in the course of discussing Smith's version of social constructivism, this settles nothing of importance.

However, it is Warren's comments on (3) which are of main interest here, for it is with respect to this point that she specifically addresses the topic of this book. She makes the following claim:

There are . . . practical implications for those who hunt if (3) is true. For example, many (perhaps all) contexts of recreational hunting, hunting for profit, and trade in wildlife will turn out to be unjustified. Consider relevant data: Hunting for sport and profit . . . is a threat to the survival of many species of animals . . . This global "trophy" hunting and traffic in wildlife occurs in economic Up–Down contexts whereby rich buyers . . . provide the demand for animals supplied by a trafficking in wildlife by sellers in poor countries. It also occurs in ecological contexts where the actions of humans threaten both the survival of species and the health (well-being, flourishing) of ecosystems. Since it is wrong to perpetuate these unjustified Up–Down systems, these practices in these contexts are wrong.

(Warren 2000: 139–40)

The factual claims involved in this passage are undoubtedly correct, but the moral judgement reached in the conclusion is one with which it is possible to take serious issue. For the hunting of animals to extinction is arguably prima facie wrong even if Up–Down contexts are not involved. Warren seems herself to be taking something like this view in her comments about 'ecological contexts' where it is the 'actions of humans' which are mentioned. Certainly the word 'context' is used in this case too, but this does not seem to be doing any real work in the case in question. To speak of 'ecological contexts' is the same as speaking of 'ecosystems'.

The point, then, is that, when one extracts from this passage the defensible moral claim made within it, it turns out to have a straightforward universal form, and the notion of context imports no further useful element – 'It is prima facie wrong to hunt animals to extinction and severely damage ecosystems.' This is in general the problem with contextualism when offered as an alternative to universalism in moral argument – it turns out to be impossible to avoid making claims which are intended by their proponents to have universal validity. And what is useful and defensible in contextualism should better be expressed in terms of the complexity of moral values and demands and the importance of recognizing the prima facie aspect of all moral judgements.

What next?

The last two chapters have critically examined one thoroughly worked out case for rejecting the approach to morality which will be used in the rest of this book. As we have noted, others besides Smith have argued against moral universalism and sought to contextualize moral thought in the ways that Smith's book exemplifies. This is plainly an area in which many more theorists' views, on each side of the issue, could profitably have been examined. The discussion in these chapters does not pretend to have been exhaustive. However, it has perhaps done the necessary job of at least beginning the defence of the approach to moral theorizing which the rest of the book embodies.

It is now time to turn from these second-level issues of how to think about moral matters to begin to mount an elucidation and defence of ecological justice. The next section will begin this task by considering those life-forms whose nature makes the case for their just treatment appear particularly difficult to sustain, namely non-sentient life-forms. We will begin with an examination of two other theorists whose general positions are, like Smith's, ones which a proponent of ecological justice will find congenial, even though, as has been the case with respect to Smith, we find ourselves taking serious issue with some of the important claims they make – namely David DeGrazia and Peter Singer.

We will first discuss DeGrazia's arguments which demonstrate that non-human animals do possess significant moral status. We will then show that his restriction of this status to the subcategory of sentient organisms is unwarranted. We will consider some further arguments on this point put forward by Peter Singer, and conclude that these arguments also fail to demonstrate that moral concern for the non-human should be restricted to the sentient. This will lead to a consideration of two recent theories of moral status which do accord the non-sentient some moral status, but on rather different theoretical bases. These are the theories of Mary Anne Warren and Jon Wetlesen. Consideration of their theories will bring into further focus issues of meta-ethics which will require resolution in the course of the development of a theory of ecological justice. Neither Warren nor Wetlesen produce a theory of justice concerning the non-human, so that we will need to extend their moral conclusions in this direction.

Part II

The case for the moral considerability of all organisms

4 The restriction of moral status to sentient organisms

In this chapter we turn to consider the first substantive issue that the case for ecological justice has to address, which concerns whether or not the granting of moral status to non-human organisms should be restricted only to that subsection of them which possess sentience. Ironically, some of the most persuasive philosophers who offer arguments for granting moral considerability to sentient non-humans are also among the most resolute in defending the view that sentience marks the limit of moral considerability. Two such authors are David DeGrazia and Peter Singer.

DeGrazia's case for the moral considerability of sentient non-humans

David DeGrazia's book does an excellent job of arguing that all sentient organisms matter, morally speaking. He pleads for a 'principle of equal consideration' extended to non-human animals (DeGrazia 1996: 44–74), while at pains to emphasize that such a principle 'does *not* entail (1) identical rights for humans and animals, (2) a moral requirement to treat humans and animals equally, or (3) the absence of any morally interesting differences between animals and humans' (ibid.: 37–8). His arguments take full account of the most powerful earlier theories in this area, including the well-known and influential views of Singer (1990) and Regan (1983), and conceivably do a much more thorough job than do those authors of grounding the case for the moral considerability of sentient organisms in a detailed analysis of those organisms' mentality. His meta-ethical position is that of Rawlsian reflective equilibrium (DeGrazia 1996: 12–14), which has the great advantage of avoiding the difficulties of foundationalism in ethics by locating moral thought within a holistic framework, but, as we will discover, has the disadvantage that some moral judgements are rejected on somewhat vague grounds, such as their not being readily fitted into the total view.

There is no space here to offer a detailed exposition of all DeGrazia's arguments. However, it will be useful to outline those which have a direct bearing on the topic of ecological justice. Firstly, the all-important 'principle of equal consideration' for non-human animals is intended to rule out a 'general discounting of animals' interests' (DeGrazia 1996: 46). That is, for example, the counting less of a non-human animal's pain as compared with the identical pain suffered by a human being, just because it is a non-human which suffers it. It also rules out, he tells us, 'the routine overriding of animals' interests in the name of human benefit' (ibid.: 47), which he interprets as involving the adoption of a purely instrumental view of animals. Finally, the principle requires that we should give equal weight to 'relevantly similar interests' (ibid.: 47–8), a caveat needed to take account of the fact that different beings often have very different interests. We will need to take further account shortly of what precisely DeGrazia understands by the term 'interests'.

DeGrazia argues that, once we accept that non-human animals have interests, then the burden of proof starts to shift to those who would hold that we have no moral responsibilities towards non-human animals. This is because universaliz-ability, the widely accepted and time-hallowed principle that like cases should be treated alike, requires that those who would reject the equal consideration principle owe us an account of the relevant differences between the human and the non-human cases which would block acceptance of that principle (DeGrazia 1996: 50–3). Failing the provision of such an account, it is more rationally defensible to accept the equal consideration principle.

Having shifted the onus of proof in this way, DeGrazia then goes on to consider various attempts to challenge the acceptability of the principle. In doing so he surveys, and in my view successfully refutes, many of the common arguments offered against according moral status to the non-human. We will here briefly summarize the arguments and DeGrazia's replies. It will be appropriate to consider some of them more fully when we seek to explore the case for ecological justice directly in Chapter 6 below.

The first challenge to the equal consideration principle is the claim that only beings capable of entering into contracts with each other can properly be accorded moral status. This mainly falls afoul of what has become known as the problem of marginal cases, or non-rational human beings.[1] These comprise a variety of cases, such as immature humans (foetuses and babies, for example) and humans suffering from certain kinds of brain damage, defect or disease. These are normally, and correctly, granted equality of moral consideration, even though they are incapable of forming contractual relations with others. The case of human beings who are 'marginal' when considered as moral agents or persons is an important topic which will recur throughout our subsequent discussion in this book (DeGrazia 1996: 53–6).

The second challenge is the sheer assertion (what DeGrazia calls the 'sui generis' view) that being human gives human beings a moral status which non-human animals lack, and DeGrazia, unsurprisingly, finds no good reason to support this view. The third is that it is the special feelings of connection, or bondedness, that give humans good reason to privilege their fellows over non-humans, in the way that we feel justified in giving more consideration to our own children over complete strangers. The truth in this, DeGrazia argues, is compati-ble with the equal consideration principle. This is because justified partiality, such as that shown towards our own children, must be differentiated from unjustified partiality, such as that shown towards our own race or sex. The former requires argument to support it, and such argument will have to recognize the force of the equal consideration principle, even when it is only human beings which we are considering. There is no reason to suppose that the force of this principle is diminished when it is non-humans which we are considering (DeGrazia 1996: 64–5).

The fourth challenge is related to the first, in pointing to a capacity – moral agency – which human beings are alleged to possess and non-humans to lack. The claim then is that only moral agents, and so only human beings, are justifiably subject to the equal consideration principle. The arguments in support of this are that (1) moral agents qua moral agents possess dignity, which is a necessary condition of equal consideration, and (2) only moral agents can reciprocate duties to each other, and if you cannot reciprocate duties, which non-humans cannot, then you cannot be owed such duties, as the principle of equal recognition requires. Why? DeGrazia considers two replies – that this is just self-evident, or that it would involve unfairness to impose duties on humans which non-humans could not reciprocate.

DeGrazia's replies to these are as follows. The 'dignity' argument is essentially question-begging. To say that human beings have 'dignity' is just another way of saying that they are appropriately subject to equal consideration. It does not pick out a separate morally relevant property. But if that is so, then the claim of dignity does nothing to show that non-humans are not also subject to the principle of equal consideration (DeGrazia 1996: 67–8). The reciprocity argument fails in connection with the marginal cases examples mentioned already. For example, we recognize moral duties to babies even while recognizing that babies, qua babies, are incapable of reciprocating those duties, and some may not even have the potential for becoming moral agents. Hence the 'fairness' argument for restricting the equal consideration principle to humans fails. For, once we admit that reciprocity is irrelevant to whether or not one is owed duties, we cannot claim that it is unfair that some individuals owe duties to beings which are incapable of reciprocating them. Concerning the 'self-evidence' claim, it is completely obscure why the pain, say, suffered by a moral agent should somehow count, morally speaking, while the same pain suffered by a sentient being which is not, and perhaps cannot be, a moral agent counts for little or nothing. (ibid.: 68–9).

DeGrazia also, correctly, argues that 'moral agency' is not a single, discrete property, possessed completely by all, and only, mature human beings. Rather it is *'a matter of degree that is not exclusively human'* (DeGrazia 1996: 70; emphasis in the original). This claim is developed in his later, illuminating, chapters where he explores how far various kinds of non-human animal may be said to possess the components of moral agency, such as 'The capacities to project into the future, to learn from experience, to keep multiple considerations in mind, to feel for others, to make decisions' (ibid.: 70). This 'gradualist thesis' effectively torpedoes the attempt to maintain the 'only human beings count morally' claim. Even where an attempt is made to cope with the marginal cases examples by finding (suspiciously ad hoc) reasons to allow marginal human agents in to the moral fold (babies are potentially moral agents; the senile are former agents) there are always going to be human beings not covered by any specific reason. If we wish, quite properly, to allow such humans to be covered by the principle of equal consideration then we cannot avoid allowing coverage to extend beyond the human (ibid.: 70–1).

The fifth challenge argues on the basis of the coherence theory of meta-ethics, or reflective equilibrium, which DeGrazia is himself supporting. It says that the rejection of the application of the equal consideration principle to the non-human, although ad hoc, is justifiable on essentially pragmatist grounds. That is, such rejection enables us to keep most of our existing moral system of thought intact, whereas acceptance of the application of the principle to non-humans would involve such a radical revision of our moral systems that they would probably be less workable than current ones. Hence, refusing to allow non-humans to count morally is a justifiable, if question-begging, price to pay for a workable moral system which we are all familiar with. No system is likely to be perfect, so the reasonable approach is to look for the best overall package, including the workability considerations just mentioned.

DeGrazia's reply is in effect to challenge the complacent assumption that our existing moral systems have a high degree of internal coherence which it would be rash to jettison in order to accept the application of the equal consideration principle to non-humans. Firstly, many people, especially those who have actually thought about the matter, do think that non-humans count morally, so that what 'our' view is on such matters is very far from settled. Secondly, the problem of marginal cases remains an unresolved thorn in the side of the 'only humans count' position. Thirdly, acceptance of the Darwinian account of human beings' and other animals' descent from a common ancestor sits ill with any thesis of a distinct human essence, and such a theory has become a key part of our total view of the world. Finally, when we actually study non-human animals and find out what their capacities actually are we find it very difficult to maintain any clear moral line of demarcation between the human and non-human (DeGrazia 1996: 72–3).

DeGrazia goes on in the central four chapters of the book to demonstrate in fascinating, and convincing, detail that the mental life of sentient animals has a complexity which, when it is recognized, makes the coherence position turn around completely. The acceptance of the equality of consideration principle to all sentient animals makes the most coherent picture of all the kinds of knowledge we take ourselves to have. Thus, what emerges from these parts of DeGrazia's book is a philosophically well-grounded and empirically well-informed case for granting equality of moral consideration to the interests of non-human animals. It is of importance to note, however, that the arguments presented by DeGrazia in favour of this position do not rely very much, if at all, on the meta-ethical basis in reflective equilibrium, or coherence, which he provides. It is possible to accept the arguments without thereby being committed to the 'coherence' meta-ethics. The importance of this will become apparent when we note that he sets a lot of store by the coherence theory in the course of rejecting the claim that non-sentient life-forms have moral status. Let us now turn to consider his arguments in this area in greater detail.

DeGrazia's arguments against the moral considerability of the non-sentient

In spite of the power and persuasiveness of DeGrazia's case, from the point of view of a theory of ecological justice his arguments have important limitations. Firstly, they are concerned exclusively with individual sentient animals, and with human moral duties with respect to such individuals. Animals are not considered in collective terms, such as being members of populations and species. This makes it very difficult to bring DeGrazia's arguments to bear on the issues with which ecological justice is concerned, such as the survival of species. It may not be impossible to argue from DeGrazia's position for some moral duties towards groups of sentient animals. For the physical and/or psychological well-being of some kinds of sentient animal may require connections of various kinds with several others of the same kind. But it may be difficult to get from such considerations to conclusions with respect to whole species.

From the specific point of view being defended in this book, this difficulty is compounded by the claim that only sentient animals count, morally speaking. It would be possible, as we noted in the first chapter, to develop a theory of ecological justice for only sentient non-humans. But we are expressly attempting here to develop a case for ecological justice as applied to all life-forms, whether sentient or non-sentient. Since the vast majority of organisms lack sentience this means that on DeGrazia's view they are beyond the reach of moral argument. No moral objection can be mounted on the basis offered by DeGrazia directly against their wholesale extermination. The best we can do is to offer the indirect argument noted in the first chapter above which bids us to refrain from their extermination on the, probably contingent, grounds that they matter, aesthetically, culturally or prudentially, to beings which do count, morally speaking. It is thus implied by DeGrazia's theory that such organisms are not directly owed any duties, and a fortiori are not the appropriate recipients of distributive justice.

Thus, putting these two points together, we find that, at the end of a closely reasoned and persuasive analysis, the practical implications of DeGrazia's theories are not particularly extensive. We are to avoid the products of factory farming, probably should become vegetarian and should reform the practices of some zoos. But there is nothing here about the need to avoid the wholesale destruction of habitats and the wholesale reform of our economic activities so as to take proper account of the morally considerable interests of other organisms, sentient or non-sentient.

Let us now focus on the arguments concerning non-sentient organisms. As we have noted, according to DeGrazia non-sentient organisms do not have interests, a view which is shared by Singer. Since moral action is directed towards maintaining proper respect for the interests of others, those beings which do not have interests cannot be the objects of moral concern. Both Singer and DeGrazia share this position too. We will consider first DeGrazia's arguments and then any additional arguments which Singer presents in support of the position.

DeGrazia early on in his book (DeGrazia 1996: 39–40) emphasizes the centrality of having interests to being the subject of ethical concern. But then he immediately begins to restrict the possibility of having interests to sentient organisms. At this stage in his argument DeGrazia does not directly come to this conclusion. Instead he introduces some key distinctions in the concept of 'having interests' which he takes to have a bearing on the matter. Following Regan (Regan 1983: 87–8) he differentiates 'preference interests' from 'welfare interests' as follows:

1   One has a preference interest when one 'has an interest' in something, in the sense of wanting, desiring, preferring or caring about that thing – that is one takes an interest in it.
2   One has a welfare interest in something when it is 'in one's interest' that one have that thing, that is it has 'a positive effect on one's good, welfare or well-being' (DeGrazia 1996: 39).

Clearly, only certain kinds of sentient organisms can meaningfully be said to have preference interests, namely ones with a fairly advanced mentality. The concept of 'welfare interests' is supposed to cover the less mentally advanced sentient organisms. However, as we will shortly discover, for DeGrazia only sentient beings can meaningfully be said to possess 'welfare interests'. This looks prima facie to be a shaky claim. After all, even a primitive organism such as a bacterium can meaningfully be said to have a 'good' or 'welfare' and thus, on DeGrazia's account, ought to be accorded welfare interests. The good of the bacterium may be said to be whatever enables it to survive and reproduce, and it may even be said to be thriving or ailing, having its welfare positively or negatively affected, in different environments as it engages in these activities with ease or difficulty. Admittedly, the possession specifically of 'well-being' may not be literally true of the bacterium, even when it is thriving, insofar as well-being may refer to the condition of thriving as experienced by an organism, which of course presupposes sentience. However, it is not clear that the possession of well-being by an organism, or the fact that it is in a state of well-being, necessarily implies awareness of that state by its possessor. After all, even in the human case, the phenomenon of hypochondria shows that objective well-being may not be accompanied by a concomitant sense of being in that state.

To return to DeGrazia's arguments, he notes that it is convenient to use the term 'interests' to cover both preference and welfare interests and to speak of 'having interests' in the case of either sort, not just preference interests. He does not, however, use any version of the argument put forward in the previous paragraph to conclude that, since having interests is a sufficient condition of being an object of moral concern, and this covers welfare interests which even non-sentient beings possess, even such beings are the object of moral concern. The reasons for rejecting such a conclusion are not given until much later in the book (DeGrazia 1996: 226–8). He there argues that 'non-sentient animals, by definition, cannot feel anything, so they cannot have aversive states. Nor can they have any other experience. Nothing matters to them; they care about nothing. They have no concerns or desires' (ibid.: 227). This emphasis on aversive states, caring, having things matter to one, and so on, rather looks to presuppose the claim that all sentient animals, not just some, have the capacity for some level of preference interest, not simply welfare interests, and that it is in virtue of this that only they count, morally speaking. For, on the face of it, welfare interests, as we noted above with respect to the bacterium example, do not appear to require sentience. To avoid this conclusion DeGrazia has to show that non-sentients do not possess welfare interests either, and as we will see this is precisely what he tries to do.

An immediate response to DeGrazia's argument is to say that, even if nothing matters to non-sentient creatures, it does not follow that nothing matters with respect to them. It matters for them that they survive and reproduce. Compare this with one of the marginal human cases already discussed – a human foetus throughout its non-sentient phase. Nothing matters to this either, but we certainly do not conclude that moral agents have no moral duties with respect to it; indeed, some would argue that it possesses many of the basic rights of a fully developed human person, even if we know for certain it is going to die before leaving the non-sentient phase, and so it does not even count as a potential sentient.

Further, even if non-sentient creatures cannot have (because they cannot experience) aversive states (feeling pain, for example) they are certainly capable of engaging in aversive behaviour, such as turning away, unconsciously, from sources of harm. Both of these points warrant the claim that even non-sentients may meaningfully be said to have welfare interests, even if they do not have preference interests.

DeGrazia admits that it is meaningful to speak of harming or benefiting a non-sentient organism, but refuses to agree that we can speak of its welfare, and thus denies that in these cases 'harm' or 'benefit' can be construed in any sense 'relevant to morality' (DeGrazia 1996: 227). His argument at this point is less clearly spelled out than usual, but appears to be that sentience (actual, and possibly former or potential as well) is a necessary and sufficient condition for an organism to possess preference interests – at the minimum, it can be in an aversive state. The ability to possess preference interests is then a necessary, and presumably sufficient, condition of possessing welfare interests – having a good, well-being or welfare. For, if you cannot have even a rudimentary preference interest (be in an aversive state, having something mattering to you) you cannot have a welfare interest either (possess a good, well-being or welfare). The possession of preference interests is thus necessary and sufficient for an organism to be an object of moral concern. Non-sentient organisms thus are not proper objects of moral concern.

DeGrazia notes that there are 'recalcitrant intuitions' that point in the other direction, towards the view that non-sentient organisms may count morally. It is at this point that his meta-ethic of reflective equilibrium is brought in to eject such intuitions from the most tenable total moral view. We should abandon the intuitions, he says, because they have no satisfactory theoretical support and our main ethical convictions are better accounted for by the sentience requirement.

To vindicate the claim that such views have no satisfactory theoretical support DeGrazia turns to consider such theorists as Robin Attfield, who argue that all and only living beings, both sentient and non-sentient, have a 'good of their own' and therefore have interests (Attfield 1991). He points out, correctly, that Attfield goes on to suggest that the interests of non-sentient creatures are 'less weighty' than those of sentient ones. He sees this as resorting to a desperate measure to conceal a fundamentally untenable view, and suggests that pressure should be put on it (DeGrazia 1996: 229). But, of course, Attfield and others can (and I think should) put pressure in the other direction and argue that something is amiss with the 'only sentients count morally' school if we do have the 'recalcitrant intuitions' already alluded to. For that school recognizes these intuitions as having some force, which is why they see the need to deal with them by appeal to the overall coherence of our moral views, rather than simply reject them outright.

DeGrazia appeals to intuitions himself when he pursues the argument against Attfield (and others who ascribe moral status to non-sentients) by assimilating the destruction of non-sentient organisms, such as pulling up weeds, to the destruction of non-organisms, such as breaking rocks. He claims that such cases 'generally do not seem even prima facie wrong' (DeGrazia 1996: 228). However, one might argue the opposite with at least as much plausibility. There is at least a moral question to be raised about the living organisms. For they are harmed by the act of destruction – specifically, they are killed. Non-organisms, such as rocks, are not harmed and, of course, cannot be killed. They are not self-preserving, homeostatic units which act so as to try to preserve their continued integrity and existence and reproduce themselves. A rock is just an arbitrarily formed piece of a larger, equally arbitrarily formed, entity. Hence, one might argue, the destruction of any organism is potentially a matter of moral concern with respect to the organism in question. We may owe the organism the degree of respect which requires us not to kill it without good overriding reason. By contrast, in the case of non-living entities, such as rocks, we can only at best have moral obligations with respect to them (they are a much-loved landmark, perhaps), not moral obligations towards them.

Another tack which DeGrazia takes (as have others) is a *reductio ad absurdum* argument which assimilates non-sentient organisms to human artefacts, such as buildings and cars. These can be damaged in ways which resemble the harms that can be visited on non-sentient organisms – their functions can be impaired, their integrity compromised. They have a complex internal structure which determines their overall functioning, unlike rocks. DeGrazia says: 'It is unclear to me why cars and buildings lack a good of their own, if having one does not require sentience' (DeGrazia 1996: 229). The implication then is, of course, that on this view cars and buildings should be accorded moral status – a clear *reductio ad absurdum*.

But the answer to this is clear enough, and is one which, at an earlier point, DeGrazia notes Attfield makes. He cites Attfield's claim that 'all living creatures have a general interest in flourishing after their kind by developing their specific capacities' (DeGrazia 1996: 228; Attfield 1991: 168). This is correct, and immediately identifies the difference between living creatures and human artefacts, for the latter do not 'flourish after their kind' or 'develop their own specific capacities' and so cannot have an interest in those things. If human beings were to create entities of which these could be said then they would have created life, not machines, and moral issues would immediately be raised.

A further argument is then put forward by DeGrazia, which is to claim that such talk of 'flourishing after their kind' and 'developing their specific capacities' implies a 'natural kind metaphysic' and a 'perfectionist value theory'. DeGrazia then argues against these positions, specifically against the 'natural kind metaphysic' by suggesting that if there are natural kinds for living things then there are natural kinds for non-living objects, such as canyons and mountains. If so, then non-living things should also be accorded interests and moral status – clearly, another *reductio* (DeGrazia 1996: 230). This argument, however, appears to involve a non sequitur. The kinds of living things are all capable of flourishing and developing, as opposed to merely changing. The kinds of non-living thing are not. So the fact that we can speak of non-living natural kinds is irrelevant to Attfield's point. It is the fact that living things are kinds of thing with certain capacities that is important.

With respect to the 'perfectionist value theory' DeGrazia offers the following criticisms. He suggests that the idea of a 'kind' is particularly problematic in the case of human beings. For them, flourishing has to be interpreted in a basically subjective way – we each autonomously decide what counts as flourishing for the particular being each of us is. So, in the case of human beings, at any rate, we cannot speak of 'flourishing after their kind' at all.

With respect to all species, including human beings, DeGrazia puts forward the view that, for reasons to do with evolution, they do not constitute 'unique natural kinds'. Species, he tells us, evolve gradually (DeGrazia 1996: 230). They do not, therefore, jump from one kind to another. If we try to cope with this point by suggesting that there is a separate 'kind' for each mutation, then all species will have to be deemed to comprise countless natural kinds, which amounts to the denial that there are any natural kinds at all. If there are no kinds then there is no possibility of 'flourishing after its kind' for any creature.

These points, however, are certainly not uncontroversial and seem to involve some exaggeration to achieve their effect. Firstly, as regards human beings, it is correct that the concept of flourishing with respect to any given individual will need to make reference to determinate forms of determinable properties. For example, reference may need to be made to the particular talents and capacities which differ between individuals and the development of which forms part of what it is for such individuals to flourish. But this does not preclude a useful statement being made about what are the determinable properties which constitute human flourishing. Some may wish to reject even this as representing an inadmissible essentialism in the discussion of human beings. But that is a point of view which is highly controversial and cannot be relied upon to mount a simple refutation of a moral point of view.

With respect to the evolution-based arguments against natural kinds, it may pertinently be replied that we can and do pick out species which are treated as natural kinds even if this is a concept with an in-built fuzziness. We do not need sharp concepts here. In particular, the claim that living organisms have a good of their own is concerned with capacities which they are likely very largely to share with other individuals. In the case of species which reproduce sexually, such individuals are those with which they can interbreed and produce fertile offspring, even if it cannot be determined with complete precision who those other individuals are. Even with non-sexually reproducing species the concept of species is precise enough to do useful work both biologically and ethically.

Further, the talk of a 'good' here does not have to be treated as perfectionist in any strong sense. Presumably, in any case, DeGrazia believes that sentient organisms can have a 'good of their own', since he thinks they may possess 'welfare' interests (which, to recollect what this means, 'have a positive effect on a creature's welfare, good and well-being') and seems to differentiate them from human beings insofar as there is in their case no 'subjective' element (they do not possess the requisite capacity for autonomy). This suggests that what is for the good of one of the higher mammals, say, is likely, *ceteris paribus*, to be for the good of others of the same kind (which may mean species, but may encompass other classifications, such as genus, family, and so on).

DeGrazia, in the course of these evolution-based criticisms of the perfectionist value theory, is taking too individualist a view of 'organism'. Evolution does create closely similar, interacting organisms which seek to reproduce themselves. This does warrant talk of their possessing a 'good of their own' not present in the case of artefacts. But these points do not imply either a rigid, clearly demarcated concept of kind or a perfectionist value theory.

DeGrazia is at his least convincing when he tries to limit moral considerability to sentient organisms. He notes that we do recognize moral demands, albeit of a rather modest kind, towards, and not simply with respect to, all living beings. His attempts to deal with Attfield and others who try to build this directly into an ethical theory miss their mark, and he becomes uncharacteristically slippery and off the point. The use of 'reflective equilibrium' to sweep away the 'non-sentients count morally' case is too blunt an instrument to do the job. The method, even on his excellent account of it, remains rather vague at crucial points such as this.

What we may conclude from this discussion of DeGrazia is that the excellent work which he does to establish that non-human creatures possessing sentience have interests, and that morality, concerned as it is with respecting interests wherever they are to be found, thus applies to them too, may be extended to non-sentient life-forms. They too have interests – welfare interests, which, contra DeGrazia, do not presuppose preference interests. Living non-sentients, which I will henceforth refer to as the 'merely living', cannot properly be assimilated either to natural non-living entities, such as stones and streams, or to human artefacts, such as cars and buildings. The moral line encompasses all living beings, not merely the sentient ones.

Singer's arguments against the moral considerability of the non-sentient

Let us now turn to Peter Singer's arguments for restricting the sphere of moral considerability to the sentient. Like DeGrazia, he does not simply reject outright the claim that non-sentient organisms count, morally speaking. His approach is the more cautious one of suggesting that 'to extend an ethic in a plausible way beyond sentient beings is a difficult task' (Kuhse 2002: 317): difficult, but perhaps not impossible. What, then, are the difficulties? The first point Singer makes is to note that it is impossible to know what it might be like to be a non-sentient creature, not for contingent reasons concerning our lack of imagination or suitable experiences, but for the simple and conclusive reason that there is nothing which it is like to be such a creature. This, he argues, means that it is impossible to say meaningfully of non-sentient organisms that they have 'wants and desires. They prefer some states to others' (Kuhse 2002: 317). He apparently accepts that it nevertheless makes sense to say of non-sentient organisms that they have a 'good or bad' and can be meaningfully be said to flourish in some situations and not in others.

However, Singer does not conclude that we can therefore use these facts to give substance to the idea that non-sentients are entitled to moral considerability. Rather, he sees three serious, perhaps conclusive, objections to the granting of moral considerability to non-sentients which derive from the lack of wants, desires and preferences among them. The first is that it is impossible to make any reasonable judgements concerning values when it is non-sentients that we are considering. Thus, we have no reasonable way of deciding that the flourishing of one kind of non-sentient organism is more important, or ought to count more in the course of moral deliberation, than that of any other non-sentient organism. By contrast, Singer suggests, we can calculate the effects on the interests of sentient organisms of different courses of action, and so can at least in principle work out how to act with respect to their interests.

This argument, however, does not succeed. There is no reason to say that it is easier to reach a reasonable view about what course of action will affect interests when we consider sentient organisms than when we consider non-sentient organisms. It may be very problematic in practice to reach a view about what sentient animals possess as preferences or wants. We cannot ask them, for none of them possess a language in which they can tell us. Our judgement concerning their preferences may be based on well-founded experiments, but it may not be possible to perform a clear and unambiguous test to enable us to determine preferences. Contrariwise, non-sentient organisms may have rather clear and unambiguous grounds for flourishing and be subject to rather obvious forms of harm and damage. It is true that we human moral actors will sometimes, perhaps often, face difficulties in deciding which non-sentient organisms possess a greater moral claim on us than others which we can adversely affect, but that seems no less true of sentient creatures. It is far from obvious that we can more easily decide which form and degree of pain, say, ought to count for more in the case of two or more sentient creatures than decide which form of harm to the flourishing of non-sentient organisms counts for more.

There is, obviously, an important issue here, which is how to determine how to make reasonable comparisons between the effects of actions on the well-being of different individuals and of different kinds of organism. Later on we will have to consider how ecological justice, which aims to secure fair shares of environmental resources to all living organisms, deals with the issue of weighing up the claims of different individuals and kinds of creature to environmental resources. But there is no reason to suppose that those issues are made more difficult if we include non-sentient organisms in the calculation.

The second objection comes rather close to DeGrazia's attempt to assimilate non-sentient organisms to non-organisms, such as rocks and rivers, and/or to human artefacts. Thus, Singer asks, 'Would it really be worse to cut down an old tree than to destroy a beautiful stalactite that has taken even longer to grow? On what grounds could such a judgement be made?' (Kuhse 2002: 318). But of course there are all the differences between the two cases which we noted earlier when discussing DeGrazia's similar arguments. The tree has a good and the capacity for flourishing, it defends itself against attack, it reproduces itself, it is living. Hence it makes sense to speak of doing wrong to the tree. None of this applies to the stalactite (whose beauty or otherwise is irrelevant to the issue, which is supposed to be a moral, not an aesthetic, one). It is possible to do moral wrong by destroying a stalactite (or river, or mountain), but the wrong done is done *with respect* to the object in question, not actually *to* the object in question. One may destroy a beautiful object or attack an important habitat or cultural entity, thereby wronging human beings and other living beings. But one does not wrong the object in question. Non-living objects have no moral claims on moral agents. Living beings do.

This is not to say that it is always wrong to destroy a living being. In the case of Singer's example, it may be easy to justify the destruction of an old tree and difficult to justify the destruction of a beautiful stalactite. The tree may be old, but not rare, near the end of the usual life-span for trees of that type, devoid of any important instrumental value for human beings or other creatures and standing in the way of the securing of important benefits for some other group of organisms. The stalactite may be extremely beautiful, formed in a rare way, making it scientifically valuable, and possess great cultural significance, and its destruction may have no positive benefits for any interests of living creatures, in which case its destruction may be very hard to justify. But the grounds of judgement in each case are very different. The tree has a prima facie moral claim on moral agents, which can be defeated by finding more weighty moral claims. The stalactite has no moral claim upon us, although many morally considerable creatures may have a moral claim upon us with respect to the stalactite.

The third objection which Singer puts forward against according moral considerability to non-sentient organisms is to claim that the attribution to a non-sentient organism of a 'will to live', or its seeking after the wherewithal for life, or the pursuit of its own good, all involve metaphorical expressions. As he puts the point:

> Once we stop, however, to reflect on the fact that plants are not conscious and cannot engage in any intentional behaviour, it is clear that all this language is metaphorical; one might just as well say that a river is pursuing its own good and striving to reach the sea, or that the 'good' of a guided missile is to blow itself up along with its target.

> (Kuhse 2002: 318–19)

To this point he adds the further claim that the lives of non-sentient organisms can be given a 'purely physical explanation' (ibid.: 319) for which we need have no more moral concern than we do for other purely physical processes.

However, the difference between a plant and a river is still greater than Singer is here allowing. The plant literally has a good, it literally possesses the ability to protect itself, to reproduce itself, to flourish or suffer harm. None of this applies to a river. A river which runs dry suffers no harm, though many creatures will often be harmed by its doing so. Rivers do strive to reach the sea only in metaphorical terms; plants do literally (if not intentionally) strive to reach the light as stags literally (if not intentionally) strive to find a mate during the rutting season. It is unclear what is meant by a 'purely physical explanation' in Singer's argument. If this phrase is meant to encompass biological explanation, then certainly it is true that the behaviour of non-sentient organisms will be explicable in biological terms, but this will also be true of quite a lot of the behaviour of many sentient organisms, including that of human beings. Furthermore, the behaviour in question will often be such that it requires our moral respect. For someone such as Singer who has strongly championed the application of Darwinian, and therefore of biological, explanation to universal characteristics of human beings (Singer 1999) – their human 'nature' in other words – this biological basis for the morally important behaviour of sentient beings must be recognized. But if it is then the assimilation of plants to rivers fails.

If, however, Singer is arguing that the behaviour of non-sentient organisms is to be accounted for in terms purely of physics and chemistry, as river and stalactite formation are, then it has to be said that that claim is simply incorrect. What has gone wrong here is that insufficient regard has been paid to the fact that non-sentient organisms are organisms. As living entities they non-metaphorically possess welfare interests, even if they are devoid of the capacity to become aware of that fact. It is the possession of such interests which makes it intelligible to say that they have a moral claim upon moral agents, even if they are not able to be aware of such claims. It is clearly true that harm done to non-sentients is harm of which they must be unaware, and it thus seems obvious that harm of which you cannot be aware (because you cannot be aware of anything) is not really harm at all. But this is an incorrect conclusion. To see this we can adapt one of John O'Neill's arguments. He has reasoned that you can have an intelligible interest in what happens after your death, even if no one survives death. For example, if after your death your reputation suffers severe and unrefuted criticism, then you are harmed, even if you no longer exist and so cannot even in principle be aware of the harm in question. For the life you lived has undergone a change for the worse – it has now acquired the character of being worthless, say, whereas before it was properly characterized as being valuable. Hence, your interests and your life can suffer harm, harm which it would be reasonable to take steps to avoid, by, for example, seeking to protect your reputation posthumously, even though the harm in question is something of which you cannot be aware (O'Neill 1993: 28–36).

Similarly, many people would argue that it is possible prima facie to visit morally objectionable harm upon a foetus by aborting it, even if the foetus which is so aborted has no capacity to become aware of the harm it suffers. These examples, then, whatever the rights and wrongs of the moral arguments in any particular case, show that it at least makes sense to speak of forms of harm to interests of which the possessor cannot even in principle be aware. If this makes sense, then it must also make sense to attribute the possibility of doing such harms to non-sentient organisms, and thus to give moral agents reasons for not engaging in such harm without countervailing moral reason.

We can conclude, then, that neither DeGrazia nor Singer have given us good reason to withhold moral considerability from non-sentient organisms. It is certainly true that there are certain harms which cannot be visited upon the welfare interests of non-sentients, which is important. We cannot cause them psychological or physical pain, so that damaging their bodily tissues cannot ever amount to torture, and we need not worry about devizing humane ways of killing them. But they can be harmed (and not just damaged) insofar as we can impede their ability to grow and reproduce themselves. They can be maimed, stunted, rendered sickly and exterminated. They share important characteristics with sentient living beings, and sentient living beings themselves do not all possess the same capacities to the same degree. There is rather a continuum of capacities across the spectrum of living things, one which is arguably sufficient to justify drawing the moral boundary around living entities rather than solely sentient ones.

If one is attracted to utilitarianism as a moral theory, as Singer professedly is, then the sentience boundary will immediately appear to be compelling. But there are very strong arguments against utilitarianism, even if one restricts moral considerability to sentient, or even just to rational, beings (see the arguments provided by Bernard Williams in Smart and Williams 1973: 77–150). It is also true that if one extends moral considerability to the non-sentient one might appear to be placing oneself in an impossible position. For moral agents are themselves organisms, and they have to eat to survive. We can be vegans, or at least vegetarians, which, if we restrict moral considerability to the sentient, appears to raise no, or few, moral issues. Both DeGrazia and Singer argue for the vegetarian alternative. But if even plants count morally we seem to be placed in unavoidable moral dilemmas.

However, these apparent difficulties emerge only if we grant equality of moral standing to all morally considerable organisms. That position, however, has not yet been demonstrated to be defensible or unavoidable. The standpoint of this book is that all living organisms do, qua living entities, possess moral standing. But this is not a simple phenomenon. We need to engage in a lot of refinement of the claim before it becomes fully defensible. It will be necessary to hold that, in certain circumstances, the moral claims of non-sentient organisms such as plants can trump the moral claims of sentient organisms, including human beings. But it will not be necessary to argue that it is always morally wrong to kill or eat other organisms, even in the case of some sentient ones. It will be important to reject a position of complete moral equality between all morally considerable entities. To some this will appear to be unacceptable. However, it can be shown that, even if we restrict our focus to the case of human beings alone, we need to, and can, justify granting more weight to some individuals and some interests than to others. In other words, no new matters of principle are introduced into the moral debate by the introduction of non-humans, sentient or otherwise, into the realm of morality.

However, these are matters which we will need to consider more fully when we have gone on to consider the arguments of some other environmental ethicists who, unlike DeGrazia and Singer, do allow that non-sentient organisms possess moral considerability. We will consider two versions of such arguments in the next chapter.

5 The moral status of the non-sentient

Although DeGrazia's theory, when it is altered so as to admit the non-sentient into the realm of morality, provides a very compelling set of arguments which could underpin a theory of ecological justice, it is not the only possible approach. In this chapter we will be looking at the theories of two thinkers who, unlike DeGrazia, are prepared to admit that the non-sentient has some moral standing, and who offer alternative meta-ethical perspectives to that of DeGrazia.

The latter's approach, as we noted, involves the concept of 'reflective equilibrium' associated with Rawls's method of moral theorizing. Mary Anne Warren, to whom we turn next, offers a pluralist approach, and Jon Wetlesen a casuistic one. We have seen some of the problems of reflective equilibrium, although the force of DeGrazia's arguments is not really diminished by the problems with the theory. Let us see how each of these other two theories fares.

Mary Ann Warren's theory

To the question of what it is that confers moral status upon any entity, Warren gives a complicated answer. In sum, she argues that there is no single property which all entities must possess in order to be the bearer of some degree of moral status (Warren 1997: 146–7). Attempts to produce such accounts always exclude entities which moral common sense (of humanity) would wish to see included (ibid.: 21–2). Rather, the truth of the matter of moral status has to be put together out of the elements of truth which such unitary theories of moral status have succeeded in grasping (ibid.: ch. 6).

A crucial distinction which structures Warren's account is that between:

1   intrinsic properties, such as being alive; possessing consciousness; possessing sentience (a property different from consciousness); being the 'subject of a life'; possessing the capacity for rational agency; and
2   relational properties, such as being part of an ecosystem; being an important part of an ecosystem (a 'keystone' species, for example); being part of a human community; being the object of caring concern by moral agents; and being an endangered species (Warren 1997: 122–3).

This means that moral status of some degree may be attributed to an entity on the basis of its possession of either an intrinsic property or properties, or a relational property or properties, or both. The possession of moral status to any degree is a matter of having a justifiable claim to the acts or omissions of moral agents. The latter thus may meaningfully be said to have certain obligations towards the possessors of such properties.

Crucially, this means that a capacity for moral agency is not a necessary condition of possession of moral status. This involves the rejection of the 'only persons count morally' view of morality, and thus allows scope for the development of a theory of ecological justice. However, Warren does not herself develop such a theory, although she does not reject the possibility of such a theory either, at least in the book from which I am drawing this account of her views.

A further point crucial to the structure of Warren's account is that the possession of one or other of two properties, one intrinsic and one relational, is sufficient for the possession of the highest degree of moral status (Warren 1997: 183). The intrinsic property is the capacity for moral agency (personhood in its fullest sense). The relational property is that of being a member of the human community, which confers the highest moral status on human beings who are not yet, or currently, or any longer, moral agents (the new-born, the comatose, the senile, and so on). This is Warren's attempt to explain why the 'marginal' cases of human beings noted in the last chapter should be accorded the full moral status of persons.

However, neither of these properties is regarded by Warren as necessary for the highest moral status. There can be moral agents who are not human beings (intelligent aliens, perhaps some of our fellow primates and higher mammals) and there can be, as we have already frequently noted, human beings who are not moral agents, temporarily or permanently. The relational property is supposed to enable Warren to reject the charge that conferring full moral status on human beings who are not moral agents is speciesism, especially if we withhold that status from creatures which are of comparable levels of intellectual attainment to the humans who are not moral agents (Warren 1997: 181). Certainly, the 'relational' claim argument does not fall foul of DeGrazia's arguments against the 'human solidarity' position noted in the last chapter, for it does not rely on any facts about human beings' feelings towards each other. On the other hand, it is less than obvious why this relational property on its own should count as any sort of basis for conferring moral status, especially as the notion of 'community' here is a rather attenuated one. If it means moral community, then, of course it does not refer to human beings alone, as Warren is herself arguing. If it means 'human species', then it looks to be a simple case of speciesism after all.

It is important to note that Warren's account depends throughout on the claim that moral status is something which can admit of degree (Warren 1997: 153). This is another sticking point for many traditional ideas of morality, according to which membership of the moral community is an all or nothing affair. If you are in, then you are a full member. There are no second- or third-class members. For Warren this is not so. Different kinds of entity possess different degrees of moral status and thus can exact different kinds of obligation from those who are full members. The club rules in this regard are set out in seven principles which assign degrees of status on the basis of the various intrinsic and relational properties which her theory recognizes (on the basis that they accord with 'moral common sense') (ibid.: ch. 6; 182–3).

The basic idea of the principles is that they enjoin upon moral agents the necessity to take account of certain kinds of interest which entities have in virtue of their possession of the moral-status conferring properties. Thus, your being a living organism entitles you to the minimal degree of moral status and requires that moral agents take full account of your interest in continuing to live fully in the manner natural to your kind. We should note in passing that Warren here speaks of 'interests' and of 'what is natural to a kind' in ways which we saw DeGrazia to find objectionable. However, it will be recollected that DeGrazia's own arguments to these ways of talking could be overcome.

For Warren, then, if you are a living organism you may not be harmed, nor may your life be ended gratuitously, without good moral reason (Warren 1997: 149–52). This will typically involve an entity with higher moral status having a claim on moral agents' actions which can only be met by visiting such harm upon you. An entity of the lowest moral status (an organism which is merely alive) can ascend the ladder of moral status if it is endangered and/or a keystone species.

The basis for this claim is not very clearly set out by Warren, and appears to rest simply upon an appeal to common-sense moral beliefs. As we saw in the above discussion of the ideas of DeGrazia and Singer, there is clearly room for scepticism here. What, it might be urged, is so morally significant about being merely alive? Are not some 'merely living' organisms, such as the smallpox virus, totally devoid of any moral significance? Our examination of DeGrazia's theory has revealed that an adequate grounding for the moral claims which Warren attributes to common sense will need some extended examination of the idea of welfare interests of non-sentient organisms. We will return to this issue when we consider the ideas of Jon Wetlesen.

Returning to Warren's argument, she goes on to assert that being a sentient organism, with the capacity to suffer pain, enables the claim to be made on your behalf that moral agents have the obligation not to inflict such suffering without good moral reason (Warren 1997: 152–6). She argues that the precise working out of what the principles require or permit moral agents to do or omit in any given instance is not itself something which can be determined by any single meta-principle. Rather it will be a matter of trying to balance the competing claims of affected organisms, taking into account the strength of their claims on the basis of their moral status.

Plainly, this approach, whatever the details, to the theory of moral status appears to be a promising one from the point of view of anyone trying to develop a theory of ecological justice. It admits non-humans to the moral community, even though, as Warren explicitly acknowledges, most of them are not moral agents and they are often in relations of mutual conflict with moral agents. It attributes to them claims which are right-like in their force, even if in most cases they do not have the structure of the fully fledged moral rights of moral agents. It allows that there are differences in moral weight between different kinds of morally considerable being, which is initially plausible and also appears to be necessary to make the problem of dealing in a principled way with irreconcilable conflicts of interest between them into a manageable process. All this, of course, only goes as far as bringing such beings within the scope of moral theory. It does not yet justify applying the more specific concepts of justice to them. It also does not go so far as to attribute moral status to holistic entities, such as ecosystems and species, but at least allows the possibility of the conferring of such status if the case can be shown that such phenomena are living entities. It gives ecosystems and species an importance, if not a moral status in their own right, on the basis of their instrumental importance for human life, and on the basis of human caring for them – an important relational property.

It does have drawbacks from the point of view of ecological justice, however. Firstly, the theory has as its fundamental structure the list. Warren lists those intrinsic and instrumental properties which human moral common sense views as conferring moral status. But the problem with a list is that it is not usually obvious from the list alone why just those things are on it, and thus whether or not the list is complete. It is a strength of the unitary accounts, which Warren effectively criticizes, that the unitary criterion at least makes it plain why only certain things are put on the list, and thus enables a check to be made that the list is correct and complete. A pluralist account is apt to have the appearance of an arbitrary selection. Can the list be given a more fundamental rationale, or is it rather beneficial for there to be an in built open-endedness to the theory?

Some environmental ethicists such as Callicott castigate pluralist theories of ethics, by which he means those which appeal to more than one moral theory (Callicott 1999: 500–2). He does not object to more than one ethical principle being applicable to a moral situation, but argues that appeal to more than one theory involves intellectual inconsistency as well as allowing the possibility of our switching between positions in a self-serving way.

Warren's approach might be held to be open to this objection insofar as she does not rest her assessments of moral status upon a single moral theory such as utilitarianism or Kantian deontology. Rather it mixes deontological, axiological, virtue ethic and generally consequentialist considerations in the course of her pursuit of what moral common sense suggests is the moral status of different entities. However, arguably this feature gives her discussion a richness and multidimensionality which unitary approaches often lack, as is evidenced in her discussion of the moral issues surrounding abortion. This may reveal that doing full justice to the complexities of the issues requires reference to be made to various moral perspectives. Perhaps no one theory can encompass all the problems. The question of whether it is unitary or pluralist accounts of moral thinking which we should adopt is one to which we will return in the course of deliberating further, in the light of Wetlesen's theory, on the appropriate meta-ethical standpoint for a theory of ecological justice.

Another cause for concern for the theory of ecological justice concerns Warren's use of relational properties – an important contribution to the debate about what confers moral status upon entities. But the appeal which she makes at several points to the relations of certain organisms to human caring and cultural values as conferring upon them moral status looks to threaten a certain arbitrariness upon the latter. Human whim appears to be able to confer or remove moral status.

Crucially, the 'respect life' principle, while it confers the right on living organisms not to be harmed or destroyed gratuitously, does not go as far as ecological justice would require in the way of conferring upon living entities a basic right to a fair share of the environmental resources they need to survive and flourish. This may not be a difficult omission to repair, however, for it involves fleshing out the idea of 'harm' more fully, in a way which brings in ecosystems and their continued existence at a more fundamental level then Warren's discussion allows. She really brings ecosystems in only to underpin a relational property which can increase the moral status of endangered or keystone species. Ecological justice would require that the relation to ecosystems be brought in fundamentally to elucidate the idea of harm to living organisms.

What one might say as a summary conclusion with respect to Warren's theory is that, while it encompasses many of the considerations to which ecological justice will wish to appeal, it errs in giving insufficient attention and weight to what she treats as a basic but not very important (morally speaking) form of moral status – that which derives from the mere fact of being a living organism.

Warren is inclined to downplay the importance of this for two main reasons. First, because merely to be a living organism, while (she holds) involving the possession of certain basic interests, also involves being devoid of an important dimension of moral status, namely its mattering to a creature what happens to it. A creature which is alive but devoid of consciousness is one which cannot be concerned about its fate (Warren 1997: 48). Hence if it cannot matter to the creature concerned what happens to it, it should not count very much in the considerations of moral agents. We may recall that it is precisely because of its not mattering to the 'merely living' what fate befalls them that DeGrazia argues that such creatures do not have interests at all. Warren, however, correctly, does not give 'not mattering' as a reason for withholding moral status from such creatures, and continues to refer to them as having interests which are morally considerable.

The second consideration is that, since moral agents are themselves living organisms, they are bound up in ecosystems with other creatures, many of which are in the 'merely living' category. This inevitably puts moral agents into conflict with many such creatures whose interests they cannot easily avoid damaging as they seek to make their own living within the ecosystem (Warren 1997: 37). Even more tellingly, such 'merely living' entities actively attack important interests of moral agents – they make them ill or even kill them. As we have already noted, since they are not themselves moral agents, they cannot be reasoned with. Often, therefore, the only effective way to deal with their attacks is to kill them in turn.

This should not cause any feelings of guilt on the part of moral agents, therefore, for they could not reasonably act in any other way. And it does not matter to the 'merely alive' in any case if moral agents do kill them.

As we have noted, Warren nevertheless allows an increase in the moral status of the 'merely living' if they are members of an endangered species, especially if they are a keystone species (Warren 1997: 166–8). The reason for this is not to do with any change in the intrinsic properties of the beings in question, but to do rather with the relational properties which they possess. This is clear in the case of the keystone species – although for even this to enhance moral status it presumably ought to be the case that the ecosystem itself is endangered and/or of importance for the well-being of entities with higher moral status. It is less clear why just being a member of an endangered species should enhance the status of the 'merely living'. Warren makes some comments of a relational nature, concerning the need to preserve the diversity of life for future generations of moral agents (humans), but this too does not involve any alteration in the intrinsic properties of organisms.

What a theory of ecological justice will seek to do in response to these kinds of argument is to emphasize more forcefully the moral status which is possessed by the 'merely living' and to argue, concomitantly, for its possession of a rather stronger right than Warren appears to allow to environmental resources, even when it is not endangered, not a keystone species, and not involved in social relations with human beings or the focus of cultural or religious importance for them.

That is, ecological justice will argue for a moral status for the 'merely living' which rests on the property of being a living organism with, therefore, welfare interests, and which holds that this property should be given greater status than Warren's 'do not kill/harm needlessly' principle allows for. Specifically, it carries a prima facie right to a fair share of environmental resources, not simply to a sufficient share of such resources to prevent the species from becoming extinct.

Warren appears to dismiss this kind of approach in the course of her discussion of Schweitzer's respect for life theory, which she claims to be impractical for the reasons already noted – moral agents have to kill and harm countless organisms just in order to live (Warren, M 1997: 37–41). However, although that is clearly true, it does not follow that the destruction of the 'merely alive' need not occasion a moment's thought for moral agents. Certainly, if we have no choice about what we have to do to make our living in an ecosystem then there is no point in giving the matter a second's thought. But we arguably do have choices, and our ingenuity often gives us alternative ways of making our living. It may well, therefore, become a serious moral question whether what we are doing to make our living as biological creatures is unnecessarily destructive of the 'merely alive'. Once the latter are admitted into the moral club, as Warren's theory requires, the basis of their membership has to be checked carefully to ensure that they are being treated in accordance with their status. It cannot be argued that because they have the lowest status they do not really count at all. They must be treated fully in accord with that status, even if it is the lowest one.

Jon Wetlesen's theory

A more recent discussion of these issues which shows some resemblance to Warren's but which offers a pattern of reasoning that avoids some of the difficulties mentioned above is that of Jon Wetlesen (Wetlesen 1999). He too argues that all organisms, even the merely living, should be accorded moral status in the sense that moral agents are supposed to have moral duties towards them. He also argues for a gradation of such status from the merely alive to the conscious, self-conscious, person and moral agent levels. He argues that human beings qualify as possessors of the highest moral status in virtue either of their being persons, with the 'capability' for becoming moral agents, or of their being fully fledged moral agents (ibid.: 302–3). Thus, he makes a distinction between being a person and being a moral agent. All moral agents are persons, but not all persons are moral agents. Hence, he does not rely on Warren's relational argument (all human beings are part of the human community) to cover the marginal humans and grant them full moral status. The argument he offers, however, involves some avowedly speculative attributions of the 'capability' for moral agency to some marginal cases, and does not quite cover some cases which Warren's can encompass, such as that of human beings born without brains (ibid.: 302).

However, Wetlesen's casuistic approach to moral argument helps deal with the 'list' problem identified earlier. As he notes, it is not clear why, on Warren's account, the merely sentient has some minimal moral status (apart from the claim that most people are prepared to accept it as reasonable to say this), and it is not clear why there should be a gradation of moral status for non-moral agents. However, if we deal with these issues in the opposite direction from that taken by Warren, then we detect a clear rationale for both of these positions (Wetlesen 1999: 295). That is, instead of moving from the cases of the merely living towards the highest level of moral status, that of moral agents, we should start with the latter and seek to extend moral status from them to non-person organisms on the basis of analogy.

In order to motivate this approach, Wetlesen introduces some further useful elements into the discussion. Firstly, he characterizes his meta-ethical position as being that of weak cognitivism, which means that normative statements are to be interpreted as based on interpersonal agreement between discussants who seek to give each other good reasons for the statements made (Wetlesen 1999: 292–3). They are, therefore, interpreted neither as realist, with a truth value and embodying knowledge (moral realism and cognitivism), nor as purely subjective expressions of emotion or decision.

It is important to note that weak cognitivism is compatible with the attribution of intrinsic and inherent value to non-moral agents. Warren distinguishes between these as follows (Wetlesen 1999: 289–91). Something possesses intrinsic value, when it is valued for its own sake by some valuer. Something possesses inherent value when it possesses moral status value which is explained in turn in terms of moral agents having some moral duties towards it. That is, the theory is deontological in that inherent value is explained in terms of moral agents' duties, rather than vice versa.

A theory of ecological justice can happily accept this deontological approach insofar as it is necessary for the direction of reasoning to proceed on the basis of analogy from a consideration of moral agents and their duties to each other to those duties which they have with respect to non-moral agents. If the attribution of inherent value were to be given a logically prior status to that of a statement of moral agents' duties, then we would be back in the predicament of Mary Ann Warren, namely that we would be reduced to mere assertion or an appeal directly to what people find it reasonable to say. Wetlesen's approach also has the advantage that it helps to explain how the application of the concept of 'welfare interests' to the merely living is to be made, namely by analogy with the application of the concept to the human case. Thus it helps us to extend the arguments of DeGrazia, which make morality a matter of respecting interests, to the case of the merely living.

It should be noted that this deontological approach is to be distinguished from the 'relational' argument put forward by Warren. The latter relies on a contingent relation (cultural, religious, aesthetic, emotional and so on) between human beings and some kinds of non-human organism in order to confer some moral status on the latter. The deontological approach concerns necessary relations between moral agents and moral patients in virtue of the latter's moral status.

Wetlesen urges that such a meta-ethics requires a theory of recognition, in order to explain under what conditions it is reasonable for discussants to recognize that reasons offered are good reasons, and he canvasses three possible such theories (Wetlesen 1999: 293–5). The first is the Aristotelian theory of dialectic, which presupposes a finite community of people agreed on fundamental value-judgements and seeking agreement in new areas. The second is the discourse theory of Apel and Habermas, which involves a universally valid ethic of discourse, and the possibility of reasoned agreement among all free discussants at the limit (a pragmatist idea), even when they come from culturally varied communities. The third is derived from the work of Korsgaard, which embodies a theory of recognition in the process of creating one's personal identity: roughly, what seems a good reason to you as an individual is dependent upon what values and commitments you have already made in the course of developing a specific personal identity (Korsgaard 1996). Wetlesen does not come out clearly in favour of any of these, seeing some merit in each of them. How should they be assessed from the point of view of ecological justice?

Aristotelian dialectic may have its place (in the form of immanent critique) within certain delimited cultures where the possibilities of securing ecological justice are already well developed. But in the case of some cultures the necessity may be to move its presuppositions in a radically new direction, which requires exposure to radically different, ecologically sensitive modes of discourse and reasoning about value matters. This suggests that the Aristotelian theory is inferior to the discourse theory and reveals an important advance of discourse ethics over DeGrazia's favoured Rawlsian method of reflective equilibrium. The latter may be confined to the activities of one intellect. The former cannot be. To put it another way, discourse ethics gives a necessarily interpersonal version of reflective equilibrium.

Korsgaard's theory may be of some use to ecological justice theory, depending upon how individualistically it is developed. However, from the point of view of ecological justice theory it threatens to swamp the development of the case for such justice in an endless series of personal appeals to human individuals. What is viewed as 'reasonable' will usually have a personal element in it, but if we take that thought too far then the possibility of widespread public debate appears to become impossible. What is left presumably would be purely interpersonal moral debate between human beings who are fully aware of each other's specific individuality.

Therefore, from the point of view of developing a theory of ecological justice the second, discourse, theory of recognition appears to be preferable. A theory of ecological justice, being inherently global in scope, requires the possibility of universal agreement, or at least sufficiently wide agreement to allow measures designed to secure ecological justice to be put in place. A possibility, of course, is not a certainty, and widespread acceptance of such a theory of recognition, requiring as it does reasoned agreement among free discussants, produces agreement only 'at the limit'. As Marcel Wissenburg has suggested,[1] It might be urged that at any point short of such universal agreement those who remain in disagreement should not be coerced into acceptance of the majority view. This is an issue, resting at the heart of liberalism, which will require further discussion. We will return to it in Chapter 7 below, where the charge of possible authoritarianism is laid at the door of proponents of ecological justice.

As Wissenburg has also pointed out,[2] Habermas's discourse theory requires the potential for communication among the participants. This may be taken to entail that beings devoid of such potential cannot figure as the objects of moral duty according to that theory. In Wetlesen's terminology, they would be at most possessors of intrinsic value rather than inherent value. What this means, however, is that theories such as those of Habermas (including the Rawlsian style social contract theories, such as that of Brian Barry), which seek to restrict moral status only to those who are participants in reasoned debate, need to be modified to allow the interests of those who are moral patients, but not moral agents, to be taken into account. The implications of this will be discussed later in Chapter 8, when theories such as that of Barry are considered.

The process of analogical reasoning which Wetlesen envisages as taking place on the basis of this meta-ethical apparatus involves arguing that moral agents should be prepared to attribute moral status value to, and thus accept some moral duties towards, non-moral agents on the basis of the relevant resemblances between them and full-blown moral agents. This is partly an intellectual matter – noting that other organisms do in fact have these resemblances of conation, consciousness, sentience, self-consciousness and so on – and partly a matter of sympathetic identification, in the manner adumbrated by Hume and others, with other organisms insofar as they do have these properties.

The key point is that this way of approaching the problem of arguing for moral status on a casuistic basis, from paradigm instances to the more or less resembling instances, has a well-motivated character which ensures that the argument is not a matter of stringing together unconnected assertions in the manner of a list. There is a further dimension to this approach which may strengthen the graded nature of the moral status that emerges by this process, and which Wetlesen does not himself introduce. This is that the properties which it is possible to detect in non-moral agents are intelligibly given a diminishing value status by moral agents when they appear within their own consciousness.

That is, these different properties, each capable of underpinning an attribution of moral status of some degree, are intelligibly and justifiably graded as higher and lower by moral agents within themselves. To see this, we have only to ask what situation you would prefer to be in yourself, *ceteris paribus*, when you contemplate this list of possibilities. The lowest section (e) sets out properties which are generally more desirable as we move from left to right, and the overall desirability of the properties taken together decreases as one goes up the list towards (a), dropping one desirable property at a time from right to left:

(a) living, unconscious, non-sentient, non self-conscious, non-rational
(b) living, conscious, non-sentient, non self-conscious, non-rational
(c) living, conscious, sentient, non self-conscious, non-rational
(d) living, conscious, sentient, self-conscious, non-rational
(e) living, conscious, sentient, self-conscious, rational.

It is reasonable to suggest that moral agents (persons), *ceteris paribus*, would rate these from (a) lowest to (e) highest in value. Just as we would in general (ignoring the possibility of a life of intense pain and anxiety for the future, say) value our own existence less as we move from (e) to (a), so it is intelligible and reasonable to attach less value to beings for whom properties (a) to (d) are permanent states. It is also in general true that it becomes harder to sympathize with other organisms as we move from (e) to (a), although this is more properly regarded as a point about moral motivation than about the moral status of such creatures. It is also intelligible why it should appear more reasonable to weigh the interests, and thus the moral status value, of creatures more as we move from (a) to (e). In terms of interests, both welfare and preference interests become more complex as we move from (a) to (e), with preference interests not coming into play until (c) is reached. This would be so even if none of these creatures was a human being (imagine that we are trying to devise a theory of ecological justice for another planet). All these points bring out fully that the differential moral weighting of different kinds of creature is being done on the basis of morally relevant considerations. Specifically it is being done on the basis of differential characteristics of different kinds of being which have a direct bearing on their welfare and preference interests. As DeGrazia forcefully showed, the latter are what moral agents are directly concerned with in moral reasoning.

The list ignores some complications, such as the possible case where one becomes non-sentient, but all other properties remain (some people seem to be in a situation somewhat like this, insofar as they are incapable of feeling pain). It is also a very big philosophical question as to what some of these terms mean – 'rationality', for example. But it remains reasonable to suggest that rationality, however interpreted, is a property, or perhaps a set of properties, of the highest importance to their possessors.

This argument does not conclusively demonstrate the intelligibility and defensibility of the graded approach, but it suggests that it is non-arbitrary to adopt it. It shows where the value judgements we are using with respect to non-moral agents come from. Someone who denies that we should grade non-moral agents in terms of their moral status, and asserts that they are all of equal moral status, will have a mismatch or dissonance between their grading of other beings and their grading of their own states, which they can only overcome by rejecting the claim that (a) to (e) represent states of higher and lower value even in their own case. That position would require some defending.

A further compelling feature of this approach is that it reveals important aspects of the pattern of argument which it is reasonable to adopt when we are considering difficult cases of moral judgement in the purely human case. For example, if the question arises as to whether to save the life of the mother or baby in a childbirth situation, where one, but not both, can be saved, it is at least arguable that the mother, being a fully fledged human person with extensive welfare and preference interests which the baby does not have, ought, *ceteris paribus*, to be given preference.

In such extreme situations – lifeboats, scarce medical treatment, triage – where hard choices have to be made between human beings who, in general, are accorded equality of moral status, it is at least a relevant consideration to ask questions about the welfare and preference interests of the affected parties in an effort to decide what, morally speaking, one ought to do. Some cases will be clearer than others. For example, many people will give preference to the mother rather than the embryo within her womb if a choice has to be made between them. More will become hesitant as we consider the foetus, or the childbirth situation. In such cases, however, it is always relevant, even if not decisive, to consider the character possessed by the parties – such as whether the inhabitant of the womb is sentient or not.

It is important that the pattern of moral reasoning we employ when it is the purely human case we are considering does involve the weighting of the claims of human beings in various circumstances, on the basis of differential characteristics that they there possess. This removes an obvious objection one might otherwise make to the differential moral weighting we need to apply when it comes to the non-human case. For it is clear that the weighting is non-arbitrary and can be employed, albeit on a somewhat different basis, in the purely human case. Thus one cannot properly object to it as involving an arbitrary bias in favour of human beings. In fact, of course, even this reference to human beings ought to be removed in the full statement of the position, for the highest weighting is being given to full moral persons, of whatever species they happen to be members, even though on this planet it is only human beings who have the capacity to attain this full status.

One further objection to this argument needs to be considered. This is that (at least) possibilities (a) and (b) in the above list refer to situations which, from the point of view of the normally functioning human agent, are of no value at all, rather than just being of the lowest value. From the point of view of such agents, it might be argued, being merely alive is indistinguishable from being completely non-existent. Hence we have simply revealed once again the indefensibility of attaching any moral value to the merely living.

This point certainly reveals a limitation to this argument. The latter, however, still works as a method of demonstrating the non-arbitrariness of the differential weighting of moral status of organisms with different characteristics. But it goes too far in the direction of favouring the point of view of the (human) moral agent. The point about human beings who happen to be in situations described by (a) and (b) is that they cannot function as moral agents, and thus as human beings, in those situations. All the properties which make the distinctively human life possible have been removed. However, other organisms which are, as a matter of their normal circumstances of existence, in these situations are able to do all the things human beings in those circumstances cannot. They can feed themselves, protect themselves from attack, reproduce themselves and in general flourish. Hence, the case of organisms which are, constitutionally, in these lowest-graded states are rescued from the complete loss of moral status by their retention of important welfare interests. Thus, the full case for the grading argument requires the ability to view the issue of moral status from the point of view of beings other than those of moral agents, even though the latter is the appropriate perspective. It is only moral agents, after all, who can consider issues of moral status at all, and their perspective is thus unavoidably a major influence on the fundamental value-judgements involved in moral deliberation.

This argument may enable us to deal with an important issue which DeGrazia spends much time discussing, namely why we human moral agents hold tenaciously to the claim that, even if other organisms have interests which can be harmed by, for example, their being killed, losing their freedom or suffering mental or physical malfunctions, the harms in question are greater for some creatures than for others, and are greatest of all for (at least normal) human beings (DeGrazia 1996: 231–2). Unless we believe this, DeGrazia argues, then the prohibition against, say, killing non-humans will be as great as that against killing humans, which is implausible. But, he argues, it is extraordinarily difficult to defend it. Once we have accepted that non-humans have interests which can be harmed, then we have to accept that they count morally (as we have seen, he thinks this applies only to sentient animals). Hence we cannot simply argue that no non-humans count morally at all. So how should we vindicate the claim that human beings' death and suffering count more than those of non-humans?

DeGrazia considers various arguments to justify this conclusion, none of which, he shows in most cases, pass muster. However, one argument which he rejects is of interest to us, since, unlike his own position, it seeks to show why non-sentients' deaths and impairments count morally, but not as much as those of other organisms or of human beings (DeGrazia 1996: 245). This is an 'objective' value argument (objective in the sense that what is being argued to be an element of an individual's good is said to be so on a basis other than the mental states or preferences of that individual (ibid.: 216)). It is offered by Attfield (Attfield 1991: 172–6). This claims that what is wrong with killing any living creature (whether sentient or non-sentient) is that it is thereby deprived of the ability to live out its potentials. Even non-sentients have such potentials. But different organisms have different potentials, and these become of increasing value as organisms increase in complexity, with the complexity of the potentials of normal human life being the most complex, and so the most valuable. This position, DeGrazia notes, needs more defence than Attfield gives it, but he does accept that at least it is not obviously a speciesist argument (DeGrazia 1996: 245), for the most valuable potentials are not in principle restricted only to the human species, and it does have the right kind of gradation of potentials to tie in to a corresponding grading of moral weight. This, we have noted, is necessary to give the good of human beings higher standing than that of other creatures, thereby preventing the choice between killing a human being and a non-human being from being a moral toss-up, and vindicating our view that in some way it is worse to kill a chimp than to kill a fly.

The casuistic approach of Wetlesen which we have just been examining may help to support Attfield's argument. For, as we have just seen, the viewpoint of moral agents is the appropriate casuistic starting point from which to assess the value of different states of affairs with respect to different kinds of organism, provided we make the crucial addition to that perspective of noting that non-sentients' potentials (and thus their interests and their good) do not rest on the possession of sentience. As we have just noted, this is not covert speciesism, for what counts as a moral agent is not a matter of species membership. Does it amount to an arbitrary, that is indefensible, favouring of the potentials of moral agents over those of the non-moral? It would be more appropriate to claim that all moral arguments, like all arguments, have to begin somewhere, and the starting point in the moral case of how moral persons value their own characteristic potentials is the one which is the least open to objection. This is bound up with a fundamental point about all moral arguments, which is that in the last resort they are all *ad hominem* – that is, start with the basic question 'How are you going to live – can you really live like that?'

To return to Wetlesen, the elements which have here been taken from his theory represent an advance upon Warren's formulation. They also provide a more satisfactory basis for applying 'welfare interests' to the merely living than is offered by DeGrazia's reflective equilibrium approach. One may conclude, therefore, that a theory of ecological justice should ground itself in a casuistic approach based on weak cognitivism and a discourse theory of recognition. It can then argue for the attribution of some degree of moral status to all living things on the basis of their possession of welfare interests and the corresponding duties of moral agents with respect to them. Warren and Wetlesen share some scepticism towards the idea of attributing rights to non-moral agents, at least to the merely living, non-sentient ones among them, but in the case of Warren, following a line of thought put forward by Midgely, the attribution of right-like claims seems to be defensible (Warren 1997: 228; Midgely 1983: 63).

However, Wetlesen's theory is expressly an individualistic one. That is, the bearers of moral status are all individual creatures. He considers, but rejects as overly speculative, the view that holistic phenomena are possible bearers of moral status (Wetlesen 1999: 316). He takes from Warren some of her ideas about how our duties towards non-person organisms may be strengthened because of certain special relationships they enter into, such as being important to ecosystems, and forming part of human cultures and societies, although he rejects her view that this involves an enhancement of their moral status, preferring to think of it as involving an enhancement of our duties while their moral status remains unchanged (ibid.: 317).

However, like Warren, Wetlesen understands the 'avoid harm where possible' principle governing our relations with the merely living as having to do mainly with what we do with respect to individual creatures. His checklist of elements to consider when trying to decide how to act when we cannot avoid harming some other beings is useful as far as it goes (Wetlesen 1999: 317–18). But it does not contain any obvious scope for considering the idea that the merely living can be said meaningfully to have a claim on environmental resources. This idea is, in effect, that we need to go beyond the duty merely not to harm or kill unnecessarily, which is all that Warren and Wetlesen appear to countenance. In other words, the concept of 'welfare interest' in their case needs to encompass claims to a fair share of environmental resources.

Let us, then, focus our attention upon the lowest moral category in Warren's and Wetlesen's (and most other people's) view, namely the category of the 'merely living'. Ecological justice requires some reasoned basis for ascribing the enhanced welfare interests mentioned in the last paragraph to the 'merely living'. If it can find this, then it will have also found a basis for ascribing it to the conscious and sentient, but non-person, categories of living entity. DeGrazia has given us good reason for recognizing the moral status of sentients. But he does not give us any arguments for regarding them as recipients of ecological justice. So we will need to find a way of encompassing ecological justice in his kind of approach. How might a reasonable case be made to justify talk of securing for all living entities their 'fair share' of environmental resources? As we will see, the development of an answer to this involves a non-individualistic approach to the attribution of moral status when it is the merely alive that we are considering. We will then need to investigate in due course whether such an approach will work with the sentient living.

Part III

The case for ecological justice

6 The concept of ecological justice
Objections and replies

Objections in principle to the idea of ecological justice

For many political philosophers the suggestion that the concept of justice can be extended in any way to the non-human living world will meet with almost instant rejection. This is because there appear to be many objections of principle which such a concept has to meet before it even gets off the ground. Let us begin by considering these.

The main ones are as follows:

1   Issues of distributive justice among a group of beings only arise when those beings voluntarily cooperate to produce and/or preserve a set of goods, such as environmental benefits. Non-human organisms cannot be meaningfully said to cooperate voluntarily to produce environmental benefits and so the question cannot meaningfully be raised of what share of those benefits different organisms should receive.
2   Justice in distribution involves the assignment of property rights to recipients of such justice. However, it makes sense to attribute property rights only to moral persons, not to organisms devoid of personhood and even, in many cases, sentience.
3   Non-human organisms on this planet are incapable of reciprocity. They cannot restrict their behaviour vis-à-vis moral agents in return for the latter's restriction of their behaviour towards them. Since they cannot do justice it is inappropriate to regard them as recipients of it.

Each of these objections relies on the underlying assumption that justice can apply only to moral agents, whom I will refer to as 'persons'. It is a key claim of ecological justice, therefore, that this is not a defensible assumption. However, demonstrating that justice can apply to living beings which are not persons is not a matter of putting forward a simple counter-argument, rather it is going to be a matter of weaving together a more or less coherent moral vision within which the idea of distributive justice to the non-person segment of life makes clear and compelling sense.

Of course, some immediate moves are obvious, such as deployment of the argument from marginal human cases which we have already encountered in earlier chapters. To recollect the main points, this argument begins by asserting that human non-persons, such as foetuses, neonates, those suffering from severe brain damage, those with severe dementia and so on, are the proper recipients of moral concern in general and justice in all its varieties in particular. Such non-persons still possess morally considerable interests. But then, the argument goes, if human non-persons can be accorded justice in this way, then the non-human non-persons with which we share the biosphere, and which also possess such interests, cannot be excluded in principle from participating in morality and justice. This is how one might deal with argument (3) above – the inability to reciprocate moral concern, including justice, does not in itself imply that one cannot be the proper object of such concern.

As we have noted at various points already, in the course of examining the arguments of DeGrazia, Mary Ann Warren and Wetlesen, it can be replied to this that human non-persons acquire their moral status by virtue of being human beings. They are deemed to have the full moral status of normal mature human beings, even though they do not have the properties (and in some cases may never have the properties) which make them full moral persons. Certainly we human beings may well engage in such 'deemings', but the justification for doing so is elusive, and looks to be a simple matter of arbitrary preference for our own species. Being a member of the human species (or of any other species) is not the appropriate criterion for acquiring moral status. Rather what is appropriate is to have the pertinent characteristics which, as we have argued, even non-sentient living beings possess.

What of argument (1)? To begin with, let us suppose that the first part of this argument is correct, namely that issues of distributive justice among a group of beings only arise when those beings voluntarily cooperate to produce and/or preserve a set of goods. We will see soon that this claim itself is open to serious objection. Let us, however, concentrate on the second part of the argument, which claims that non-human organisms cannot be meaningfully said to cooperate voluntarily to produce environmental benefits and so the question cannot meaningfully be raised of what share of those benefits different organisms should receive.

Against this argument it can be urged that environmental benefits are assuredly produced by the joint actions of the component organisms in the biosphere in ways which we are still unravelling. The most famous version of the ecological claim on which this argument rests has been articulated by James Lovelock in his Gaia hypothesis, which conjectures that the planet's biosphere, including its non-organic components, makes up a superorganism which actively (though unconsciously) works to maintain the conditions within which life may persist in the face of disruptive external forces, such as large changes in the electromagnetic radiation produced by the sun (Lovelock 1979). This is still a controversial hypothesis, but the basic idea which it employs is not. It has become a commonplace of ecology that natural processes confer many vital environmental services for all the planet's inhabitants. There have even been attempts made to work out the monetary value of all the goods and services which are provided by the biosphere to the human economy. As Tim Radford reports, 'A team at the University of Maryland once calculated that nature delivered goods and services worth $33 trillion to the global economy every year. The gross national product of the whole world at the time was only about $18 trillion' (Radford 2003a: 5).

This production may not be voluntarily undertaken, and much is the result of the actions of organisms devoid even of consciousness. Indeed, it appears that it is the activities of single-celled organisms and the small, non-sentient multi-celled organisms which are crucial to the health of the biosphere in a way that larger multi-celled organisms are not. As Sir Robert May put the point, 'It's the little things that make the world work' (BBC 2000b). This has important implications for the claims in justice of non-sentient life-forms, which make up the overwhelming majority of the earth's species. For, it can properly be urged, it is the fact of the production of the benefits that raises the issue of justice, not the question of the conscious or voluntary nature of that production. We can see this if we note that the non-voluntary nature of the cooperative productive efforts of human slaves does not remove their claim in justice to a fair share of the products of their labour, although the institution of slavery will need to be abolished before justice can be done. Similarly, we may argue, the fact that organisms jointly produce the environmental benefits which sustain life entitles them to a fair share of those resources.

Marcel Wissenberg has offered the following reply to this argument.[1] Human slaves produce against their will, as the result of the exercise of coercion against them. But they have the capacity to undertake such production voluntarily, for their joint benefit. Non-humans are incapable of this, and so their non-voluntary productive activities are not engaged in against their will. 'Non-voluntary' in their case means 'devoid of will', not 'against their will'. They thus lack the capacity for voluntary productive activity which alone entitles them to considerations of justice in the distribution of those products.

This, however, appears not to be a conclusive rejoinder. For, if a group of human slaves were brainwashed or demoralized to the extent that they lost the capacity for voluntary cooperation there would be no diminution in the strength of their claim in justice (although this might need to be made on their behalf) to a fair share of what they had jointly produced. If so, there is no non-question-begging way of avoiding questions of fair shares with respect to the joint production of non-human organisms. Of course, working out 'fair' shares may be a very difficult task, some of the complications in which we will shortly be noting.

It might be urged in reply to this argument that, if it makes sense to speak of just or fair shares of environmental benefits for all participating organisms in the biosphere, then it must also make sense to speak of organisms as sometimes taking more than their fair share. Since we are considering non-person organisms, incapable of a sense of fairness, and thus unable to restrict their activities within agreed moral limits, this implies the necessity for persons, human beings on this planet, to act as moral police so as to restrain such unjust takings, perhaps by such measures as culling members of the offending species.

For many, the idea that human persons should act as moral police in this way will be an immediate *reductio ad absurdum* of the whole idea of ecological justice.

However, this would be too hasty a view. There are certainly good prudential reasons for human moral agents not to engage too readily in such policing, for our understanding of the nature of interactions and of the causes of harms and benefits among non-human organisms is still very rudimentary. Only in cases where a species is clearly being driven to extinction by the usurpation of its life-support systems by another culprit species in such a way that such usurpation can be prevented by identifiable and practicable human actions ought human beings to interfere, *ceteris paribus*. Where such usurpation is the result of prior human activity, as in the case of the introduction of alien species to an ecosystem, the argument for human beings to seek to restore the status quo ante are particularly strong, provided the steps can be quickly taken in a timely manner and do not risk making the situation worse. There are, then, clear practical problems with human attempts to alter the structure of ecosystems in a way designed to favour some species over others, but, in spite of these practical difficulties, there is no objection of principle to this moral policing view once one grants non-person organisms the moral status which makes their being recipients of justice an intelligible possibility.

Of course, what is emerging here is that the concept of ecological justice imposes obligations on human moral persons to act on behalf of the claims in justice of non-person organisms to their fair share of environmental resources. More radically still, the granting to non-person organisms the moral status which brings them into the community of justice as recipients, though not as dispensers, involves the moralization of the non-person world of organisms.

This moralization of the non-person world becomes inevitable once persons emerge onto the scene. It might appear at first sight to be preferable for non-person nature to continue to be treated in a non-moral way, for the interaction of biological and moral relationships between human beings and other organisms leads to some very difficult problems. But to think thus is to fail to see that the relations between moral persons alone already encompass biological issues, for the persons are themselves organisms which interrelate in biological terms. Questions of sexual relations and reproduction, of the provision of food and other biological requirements for life, are large and weighty moral issues at the heart of human existence. The intertwining of persons' lives with those of other organisms raises further difficulties, but does not introduce modes of moral debate which are not already present in the purely human case.

Only if the issues of morality could be hermetically sealed off within a world of disembodied persons could these difficulties be avoided. However, all the persons of which we have knowledge are embodied (and I have argued elsewhere that persons are necessarily embodied; see Baxter 1996) and so the biological/moral issues between them are unavoidable. We might be able to avoid the moral problems which arise between embodied moral persons and other organisms if the former could be sealed off hermetically from the latter. This might eventually be done (if, for example, we all go to live under a glass dome on the moon, meeting our biological needs by the application of technology directly to the non-organic resources there available to us), but it is not our present situation. As we have already noted, in that situation, on most conceptions of morality, the non-person world has some moral status, and some moral claim on persons. The attack by persons on the life-support systems of the non-person world is, then, what gives the concept of distributive justice its point and purchase.

It is appropriate to make this point by employing the phrase 'circumstances of justice' to refer to the state of affairs which exists between all the organisms on this planet, human and non-human. This phrase has usually been employed by philosophers to refer to some specific features of the human condition, such as relative scarcity of resources and limited human altruism, to motivate the search for some mutually accepted basis upon which groups of human beings could reasonably agree on how to divide up the scarce resources between themselves. Those circumstances, however, are now revealed to be much more complicated than allowed for in classical discussions, such as in those provided by Hume or Marx.

Certainly, human beings still can and do adversely affect each other's welfare interests through their impacts on the environmental resources to which all need access. But they can do so now across time and space in a manner not readily envisaged in previous periods. The idea of the 'ecological footprint' neatly embodies this concept. This is that individual human beings appropriate certain amounts of the planet's environmental resources to service their life-styles. Currently this averages at 2.1 hectares (5.2 acres) per person, with a wide differential across the planet's human inhabitants (Radford 2003a: 4). Some theorists (see Dobson 2004) use this concept to argue for a new understanding of justice between human beings as part of the idea of environmental citizenship. Clearly, this kind of conceptualization of justice spreads the issue of distributive justice far beyond its traditional, limited domain, and crucially allows non-human organisms to be covered. For human beings can take more than their fair share of environmental resources from all morally considerable beings, not just their fellow humans.

Thus, one might reasonably claim, human beings are in 'circumstances of justice' with respect to each other (where the 'other' now encompasses human beings in distant societies and future human beings) and with respect to other morally considerable beings with which we share the planet. Each can adversely and beneficially affect the welfare interests of the others. This is not a situation which any of the planet's inhabitants has chosen. It does not arise on the basis of contract or agreement. But this does not mean that the conditions do not pertain within which issues of distributive justice can meaningfully and pertinently be raised. It should be stressed that it is the fact of our interconnectedness which is doing the important work here. The fact that all the organisms of the planet share a common descent, share DNA and have other characteristics in common is not what creates the circumstances of justice. Even if organisms had none of these features in common, it would be their joint occupation of a biosphere from which all draw their sustenance and which all have the potential to affect adversely or otherwise that would be the crucial point.

However, this line of thought raises another concern which defenders of ecological justice must address. This is the issue of who may properly speak on behalf of non-human organisms, given that they cannot speak for themselves. What legitimizes the appointed or self-appointed representatives of the non-human world in their claim to defend the interests and/or moral claims of that world?

There are some obvious initial replies which can be made to this question. The first is that one can only be in a position to articulate the interests of non-person organisms if one can demonstrate appropriate degrees of knowledge of those organisms' biology, ethology and ecology. What 'appropriate degrees' are is itself a matter for debate. It is far from implausible, for example, that in many instances the human beings who possess such knowledge will be members of traditional societies having direct interaction with the non-humans in question, and in possession of extensive knowledge based on their traditions. But whether in a given instance it is those who have received a tradition-based training, or those who have acquired the scientific skills in the appropriate disciplines, who are the better placed to offer an account of the interests of non-humans, this, being an empirical matter, will be subject to the usual caveat of fallibilism. This is a problem, but one which is not unique to the case of the attempt to protect the interests of non-human non-person organisms. In the case of human non-persons, too, our understanding of their needs, physical or psychological, are subject to ongoing determination by the appropriate scientific disciplines. What seemed a tenable viewpoint at one time sometimes turns out to be untenable, as understanding advances. It follows, however, that when the issue of which course of action will best protect the interests of non-human non-persons is open to dispute between parties, each of which has satisfied the normal criteria for establishing appropriate knowledge, then the matter may have to be left to an impartial adjudicator to determine, perhaps on the basis of precedent or judgement.

In the case of most non-humans, however, the business of judging their interests is made a bit easier to handle by virtue of their much simpler psychology, or complete absence of it in the case of non-sentients. However, it is clear that we have still a great deal to learn about the mentality and aptitudes of other species, as well as about their physical needs, so that it should never be assumed that claims to knowledge of the needs of non-human non-persons are never open to question.

There is a check in the case of the protection of the interests of human non-persons, which is not available in the non-human case, which is that all the participants to the discussion are themselves human beings, with their own knowledge 'from the inside' of what interest claims have a background plausibility. But this, though valuable to impartial judges given the task of adjudicating between disagreeing 'experts', is not itself infallible. What is the common-sense view of such matters is as subject to revision as any other form of knowledge claim. Certainly it is not enough to make a distinction between the human and the non-human case which is sufficiently clear-cut to justify denigrating the latter as being a complete non-starter.

There are, of course, arguments of a more metaphysical nature, to the effect that human beings can never grasp the interests of non-humans, since they can never enter into the lives of those creatures 'from the inside'. But even if we grant that the inner lives of non-humans must be a closed book to us, in the way that the inner lives of other human beings are not, this does not encompass all the areas germane to the determination of harm to interests. Physical injury and death appear to be harmful to the interests of any organism which undergoes them. In any case, as we have seen, DeGrazia has offered compelling reasons for rejecting this whole line of argument with respect to those sentient organisms which are the only ones to which it has application.

If it is possible for human beings to come to an informed and defensible view of what is in the interests of non-humans, of the kind necessary to make decisions of ecological justice possible, then a further implication of this requires to be noticed. This is that the establishment of ecological justice as a part of human society's basic aims will make it necessary to foster much more extensive knowledge than is currently prevalent of other life-forms and their needs. This will be needed both as a matter of securing the requisite expert knowledge in matters of complex public policy decision-making, and as part of the basic repertoire of citizenship knowledge, to inform democratic control and criticism of decision-makers' actions. From the point of view of ecological justice, this may have the beneficial effect of providing a positive feedback mechanism working in favour of the deeper embedding of ecological justice within a society which embarks upon it. For one common cause of the failure of human beings to be willing to accept the moral standing of non-humans is simple ignorance of them. Contrariwise, the emergence of a sense of obligation to the non-human in traditional religions and other cultural forms may in large part derive from the people in those cultures having a direct daily contact with, and so knowledge of, the lives of the creatures in question.

Returning to the issue of the qualifications and bona fides of those who seek to speak on behalf of the non-human, another obvious point is that there should ideally be an absence of substantive self-interest among the would-be spokespersons in the particular outcome of the debate. This is only a desideratum rather than a requirement, because in a given instance it may be impossible to find parties on either side of a dispute concerning the interests of the non-human who do not have direct material interest in the outcome of the debate.

These points are perhaps sufficient to establish that there is no difficulty in principle in recognizing the qualifications and bona fides of at least some human beings to speak on behalf of the interests of non-humans. There may be difficulty in any specific instance in determining that a given individual or group satisfies these requirements, but these difficulties will not be peculiar to the case of non-human interests. It will undoubtedly be necessary, in a legal context, to establish rules for determining standing to contest or determine a decision regarding the interests of non-humans in a given case, and the state will undoubtedly have to empower by statute appropriate bodies to be granted standing in particular areas. But in the more general realm of ongoing public and political debate, anyone who is a moral agent has the standing necessary to take a view and offer it to others on the issue of what interests of non-humans require defence and how conflicts of interests between humans and non-humans should be resolved.

At this juncture, however, another important matter requires to be addressed.

Our consideration of the first objection has led us to the view that the issue of distributive justice with respect to environmental resources properly arises once we grasp fully the implications of the point that it is the joint activities of living organisms which contribute to the environmental benefits that all of them enjoy. We have argued that it is not necessary for the organisms in question to undertake this joint production on a conscious, contractual, basis for issues of justice to be raised by it.

The issue which this introduces is whether this point could be a sufficient basis upon which to erect a theory of ecological justice.[2] In other words, is it possible that we would need to consider only the type and amount of contributions which each organism provides to the sum of environmental benefits that all enjoy in order to determine what its proper share should be of those benefits? If it were possible to do this then it would appear to obviate the need to work out a theory of moral claims based on the possession by organisms of some moral-status conferring attributes, of the kind we have encountered in the work of DeGrazia, Warren and Wetlesen (and many other environmental ethicists).

There are reasons to suppose that this suggestion, tempting though it is, should be resisted. The first is that something's making a contribution to the sum of environmental benefits is a necessary, but not sufficient, condition of its being an appropriate recipient of ecological justice. For example, many non-living entities, such as inanimate physical phenomena – mountains, rivers, clouds, minerals, radiation and so forth – make a contribution to the sum total of environmental benefits, but are clearly not the appropriate recipients of any kind of justice, for they have no interests. Hence, we still need a theory of moral considerability to pick out from the class of contributors of environmental benefits those which are the appropriate recipients of justice.

Secondly, given the highly complex nature of the interactions which are present in ecosystems, it may be extremely difficult in practice to work out what precisely the contribution of any organism actually is, even though in some cases it may be quite easy to discover some of the major contributions which some organisms make – such as keystone species. It often looks as though some organisms can disappear without obvious loss to the overall viability of the ecosystems within which they operate. Some may be functionally redundant, in other words. However, we need to be careful about such judgements of redundancy, given our still rudimentary knowledge of ecosystems. The usual environmentalist point about threshold effects applies – such species may be a rivet holding the system together. Some rivets may be lost and yet the vessel survive, but it is a step nearer to possible collapse as a result. The main point, however, is that, given the complexity just alluded to, it will be extremely difficult to formulate a theory of ecological justice on the basis of a calculation of contributions alone.

The third point is that the aim of the argument about joint production of environmental benefits has to be kept clearly in view. The point made above about the interconnectedness of all organisms and the contribution they make to the sum of environmental benefits was designed to counter the claim that justice applies only to beings which voluntarily cooperate to produce joint benefits. This claim, it will be recollected, was designed to rule out the very possibility of counting non-humans in to the discourse of justice. We have countered this argument by showing that non-human organisms do jointly contribute to vital environmental benefits, and that the voluntary nature of the contribution is not relevant to the issue of justice. But that argument does not require us to show that all organisms contribute large or important amounts to the overall provision, or that their contribution, however large, is indispensible. It is sufficient for the purpose for which it was designed for the argument to show that non-humans do contribute enormously to the provision of vital benefits, and that thus they do gain admittance to the discourse of justice.

However, the issue of justice towards them cannot depend simply on the amount of benefits which they contribute. We can see this if we imagine the following scenario. Suppose that scientists discover that a certain species of bacterium is a crucial keystone species for the whole biosphere. If it were to disappear the biosphere would rapidly deteriorate, and they have ascertained that no other species is able to fill its ecological niche – that is, there is no redundancy operating here. If contribution to the production of overall benefits were the sole criterion employed to determine the amount of environmental benefits to be received, then this species of bacterium would presumably be entitled to the largest amount. On the other hand, it might well turn out that the human species is not a keystone species and is highly redundant. If it were to disappear from the biosphere the latter's viability would be unaffected, or perhaps even improve significantly. Thus, if the criterion of contributions to the environmental benefits were to be the sole criterion of distribution of those benefits, then the human species might be entitled to rather little.

What seems to be unsatisfactory about this is that the issue of what interests each kind of organism possesses has been left out of the picture by the 'contributions to environmental benefits' criterion of ecological justice. What is wrong with this is that it is not irrelevant to the justice of the distribution of benefits what interests are possessed by the potential recipients of such benefits. This is true even when we are considering the purely human case of the distribution of benefits which are the product of human voluntary cooperation. Which is to say that a purely contribution-based theory of distributive justice is not acceptable even where its applicability is least open to question. In the human case we do think that claims to cooperatively produced benefits are properly determined in part by the interests possessed by beings with moral standing, even if, as in the case of the new-born, the senile, the brain-damaged and so on, they are incapable of making any contribution to the production of those benefits. This is a claim to which we will return in Chapter 8.

What the point made in the last paragraph amounts to, of course, is that the first statement in objection (1) at the start of this chapter should be rejected. That is, it is not correct to assert that issues of distributive justice among a group of beings arise only when those beings voluntarily co-operate to produce and/or preserve a set of goods. The characteristics and thus the interests of the possible recipients of the benefits must also be taken into account. This, therefore, is the main reason why the theory of ecological justice cannot ignore the interests of, and thus the characteristics possessed by, different kinds of organism. A theory of ecological justice cannot avoid the attempt to distinguish the claims in justice of different kinds of organism on the basis of the different kinds of interest each possesses. This is a task to which we will have to turn when we have completed the arguments of principle for the admission of non-human organisms to the community of justice.

Let us now, however, consider argument (2) – that distributive justice involves the attribution of property rights to the recipients of justice, and only persons can meaningfully be said to hold property. If this claim rests upon the assertion that property is a social institution, constructed by persons, and that non-persons cannot participate in it, then this is a version of the argument which falls prey to the 'marginal cases' argument already mentioned. Human non-persons may meaningfully be said to have property rights even though they are incapable of understanding this concept and incapable of respecting the property rights of others.

Granted, there are rights involved in property which neither human nor non-human non-persons can meaningfully be said to exercise. Hence, the concept of property which we may use in such cases will be somewhat attenuated when compared with the ordinary human case. However, we need to recall the by now standard point that property encompasses a bundle of rights (see, for example, Becker 1977). It is certainly correct to argue that some elements of this bundle cannot be literally attributed to non-persons, human or otherwise. For example, the right to manage and to bequeath both presuppose conscious intention, which many non-person organisms simply lack. We may speak of the ants 'managing' the affairs of the colony, or of an animal 'inheriting' its parents' territory, but these seem to be metaphorical expressions. However, other elements of the bundle of property rights can straightforwardly be attributed to non-person organisms, such as the right of access and the right to use. These rights will certainly form the central core of any idea of ecological justice. Hence, an attenuated concept of property does seem applicable to non-persons.

It is to be hoped that these replies have served to vindicate the concept of ecological justice against the first three obvious objections of principle. However, we should now make some immediate concessions to those who are sceptical of the whole idea.

Firstly, the defence of ecological justice will involve the extension of the concept of 'justice' in a new way. For the rights to which the theory gives rise will be different from those attributed to moral agents in traditional theories.

This is precisely because the entities to which considerations of justice apply, the community of justice, will be broadened to contain entities which are not moral agents, and which are not even conscious in many cases. Hence, the rights which it is required to grant to moral agents will not be present in the case of ecological justice. The community of justice will also encompass entities which are not even in various kinds of relations with moral agents, relations which, as we noted in the first chapter, are sometimes used to underpin moral concern for those entities. Thus, they are not cared for, valued, admired, or the focus of cultural or religious values. Any rights which derive from these sources in the case of human beings and some non-human beings will not be present in the case of ecological justice.

More dramatically, the community of justice will now also contain entities which routinely attack and harm moral agents, not as a result of conscious malevolence, but as the result of the workings of their inner nature. As we noted in the discussion of Warren's theory of moral standing, such beings, being devoid of the requisite forms of personhood, are not to be reasoned with. No consciously undertaken modus vivendi can be struck between them and moral agents. Further, it will contain beings which are routinely attacked and harmed by moral agents. This too will usually be not as a result of malevolence, but as a matter of necessity. The moral agents in question are human beings, and thus are organic beings embedded in ecosystems from which they must draw their sustenance and which contain threats to their well-being against which they must defend themselves. Such life-preserving and life-enhancing activities inevitably involve the death and harming of countless other beings with which human beings are entwined.

This means that often the conflicts of interest between human beings and the non-human cannot be supposed to be resolvable, even in principle, by any form of reasonable compromise. Sometimes, perhaps often, the interests of one party must be wholly overridden. The rights of non-person organisms to which ecological justice gives rise, therefore, are strongly prima facie. Unlike, say, the equal rights and liberties of Rawls's theory, which can be given absolute protection in a system of lexical ordering, the rights of non-person beings in the sphere of ecological justice can all be overturned in certain circumstances.[3]

It goes with this conception of rights that an attempt must be made to work out on some principled basis a system of moral trade-offs between the competing interests of all morally considerable beings. This is because the danger for ecological justice is that it may turn issues of distributive justice into a moral toss-up, an issue which we have already noted in passing at the end of chapter 5. As we will soon discover, the attempt to produce this system of trade-offs in turn leads to the attempt to attach differential weightings to different interests. The result of this is that it will not always be the case that the interests of any moral agent will trump any interest of any non-moral agent.

For some the idea of relations of justice between entities which, to put it crudely, eat each other is an absurdity, and the stretching of the concept of a right just outlined involves its destruction. But, while the claim that justice can obtain only between members of species which are moral persons (including individual beings from that species who do not happen to be moral persons in given instances) makes the theory of distributive justice easier to handle in many ways, it falls afoul of the fact that moral status of some sort is inevitably extended by moral agents to living entities that are not moral agents. The case for ecological justice emerges ineluctably once it is recognized that the admission of the non-human to the moral sphere opens up questions concerning how to protect the interests of those beings in ways to which a theory of distributive justice is an appropriate reply. Merely refraining from causing the sentient segment of the non-human world unnecessary pain is an inadequate response to those questions. But the complexity of the interconnections of living beings, and the fact that only some of them are moral agents, takes the concept of justice in new directions. We will return at the end of chapter 8 to the issue of whether the concept of distributive justice can survive the alterations necessary to accommodate the interests of the non-human.

One further issue now needs to be addressed. So far we have attempted to show that the idea of justice towards the non-human makes sense and that the moral necessity for human beings to recognize and act upon such justice can be demonstrated, and have indicated how justice is to be understood in this context. However, we still need to say something more about human moral motivation in order to establish that it is not a forlorn hope that human beings can be brought actually to act on such a conception. This is a version of the problem which Thomas Nagel identified (Nagel 1991: 4–5), namely how to bring the universal, impersonal standpoint of morality into contact with the subjective standpoint of personal motivation. Failure to do this, Nagel argues, leaves us with, in effect, a motivational deficit. People may grasp the moral imperative intellectually, but not put it into practice because they see no reason from their personal perspective to do so. This problem, arguably, is at its strongest when it is the interests of non-humans that are at stake, for our human failure to meet our moral obligations towards them is not something which the non-human is in a position to do anything about. Of course, in human cases we can establish a system of rewards and punishments to make the personal standpoint and the universal one coincide for the morally obtuse or recalcitrant, in the hope that the universal standpoint of morality may be suitably internalized by them. But this needs at least many human beings to make the universal standpoint of morality their own to begin with.

The idea that moral behaviour towards one's fellow moral agents is in some deep way in one's interest has long been offered as part of the persuasive reasoning here. The concept of self-interest that is employed in such arguments, however, is not a narrowly egoistic or crassly material one. Rather it depends on arguing that each moral agent has legitimate interest in becoming most fully what he or she essentially is, which is, precisely, a moral agent. It involves being true to what one deeply and fundamentally is. This notion is otherwise known as integrity.

Since this self-interest argument does not depend on the idea of a quid pro quo, but of being true to one's real nature, it does not require that moral agents only concern themselves with respecting the interests of other moral agents, who are in a position to reciprocate. It thus also encompasses those beings who are only moral patients, and who are incapable of such reciprocity. Thus, once it has been recognized that there exist other morally considerable beings besides moral agents, and that moral agents are in a position to affect, for good or bad, those beings' interests, then it is in the self interest of moral agents qua moral agents to recognize and respect those interests in their actions, both individual and collective.

This is essentially Tim Hayward's argument (Hayward 1998: pt II), that when human self-interest is properly understood it can be shown to be in the interest of human beings, as a perfectly general matter, to recognize the moral considerability of the non-human and, importantly for the purposes of this book, to seek to do justice to it by acting justly towards it:

human beings have no good reason to withhold respect from nonhuman beings: and this issues in a prima facie obligation to work out what respecting them means in practice. If human beings respect one another as rational moral agents, then fulfilling the demands of this mutual self-respect requires them to view rational moral agency as a source of responsibility in relation to their cohabitants on this earth, rather than as a mere privilege.

(Hayward 1998: 119–20)

But this notion of self-interest is to be understood as distinct from egoism, and, for social and ecologically embedded creatures such as ourselves, as encompassing both a concern for the interests of others and a recognition that others partly define what counts as in one's self-interest (Hayward 1998: 75–6).

If these arguments are sound, then they serve to give an answer to those who worry that human beings' self-interest may prove to be an insuperable barrier to the pursuit of ecological justice as this book has characterized it. They are, of course, rather abstract arguments, and some may deem them too high-flown to have much practical persuasive power. But all moral argument has this quality when put in the most general terms, and there are likely to be many more arresting ways of expressing the point which will be more telling for people who are impatient of such abstractions in the realm of moral thought. In any case, if this self-interest argument is rejected the problem of moral motivation remains a serious one for even the purely human case. If it is accepted as a remedy for moral motivation deficit in the human case then, as we have seen, it is not possible to resist it in the case of the actions of humans towards non-humans.

Let us now turn to consider another pressing issue which arises when we seek to defend the concept of ecological justice. We need, in the remainder of this chapter and the next, to try to determine which, if any, of the existing approaches to the generation of a theory of distributive justice is best placed to accommodate the needs of ecological justice. There are three possibilities we need to consider – the Rawlsian version of contractualism, the analysis of distributive justice as resting on shared social meanings provided by Michael Walzer, and the analysis of justice as the outcome of the workings of the market provided by Robert Nozick. These three have the merit of providing an understanding of justice in distribution which has implications for the practices and institutions of human societies, albeit heavily influenced by the liberal societies within which they have all developed. That is, they are not simply sociological or historical discourses on the significance of practices of justice. They therefore provide some promising avenues for the development of a theory of ecological justice, rather than simply a diagnosis of the socio-cultural forces which lead people in the direction of trying to attain one.

In this chapter we will look at the ideas of Nozick and Walzer. These will not detain us for long for they turn out to be not very useful for the exponent of ecological justice. In the next chapter we will consider the ideas of Rawls more fully to ascertain how far a Rawlsian-style contract theory can accommodate the needs of a theory of ecological justice. This will require us to consider the ideas of those who have sought to turn Rawls's theory in the direction of ecological justice, as well as those who argue that this cannot be done. This will raise the more general issue, already much discussed, of how far a liberal can go in the direction of recognizing the moral considerability of the non-human.

The theories of justice of Nozick and Walzer

Walzer's theory of distributive justice among human beings, encompassed in the phrase 'spheres of justice', was, he tells us, specifically worked out as a response to the analysis of justice offered by his colleague Nozick, expressed in the latter's *Anarchy, State and Utopia* (Walzer 1983: xvii). The two approaches to the theory of justice are certainly very different, and reach very different conclusions. Nozick's starting point is with a theory of basic rights for individual human beings which are intended to act as side-constraints upon what individuals, and groups of individuals, may do to each other, morally speaking (Nozick 1974: 28–33). The recognition by individuals of each other as autonomous members of a Kantian 'kingdom of ends' requires that each be accorded a set of basic rights which erect a 'no-go' area around each of them.

According to Nozick, this need of individuals for autonomy – the ability of each of them to determine for themselves how they will live their lives – arises as a condition of each one's having a meaningful life, a matter to which each must attach supreme importance. He remarks that the idea of the meaningfulness of one's life looks to be a suitable candidate to straddle the is/ought divide (Nozick 1974: 50) – that is, from the fact that an act one is about to perform will render one's life meaningless gives one without further ado a reason to believe one ought not to perform it.

From this starting point Nozick draws his further conclusions concerning the role of the state in a society of mutually recognizing autonomous individuals. The role of the state, in a nutshell, is to preserve the system of basic rights, including the right to acquire property, from morally impermissible forms of incursion, such as theft and fraud, which individuals might otherwise be able to engage in. When individuals acquire property in ways which do not violate the rights of others, including satisfying the 'Lockean proviso' to 'leave as much and as good for others', then no moral objections can be raised to the distribution of property which eventuates, unequal as that may turn out to be (Nozick 1974: 178–82). The state is specifically debarred from engaging in enforced redistribution of holdings between citizens in the name of some abstract moral criterion of distribution of the kind much favoured in the socialist tradition. This minimal state is, he argues, the only morally permissible state when we begin from the Kantian premises which also inspire the work of Rawls (ibid.: 149).

This summary is inevitably simplified, but accurate enough for the present purpose, which is to assess this theory from the point of view of ecological justice. With respect to this issue it is immediately apparent that the theory's crucial emphasis upon autonomy, and its connection with a meaningful life, automatically debars the extension of this conception of justice to non-persons. Certainly, as we saw in the earlier discussion of DeGrazia's theory, it can be argued that at least some non-human organisms possess something like the autonomy which characterises human beings. They are at least capable of self-direction or self-formation (autopoeisis), and certain of their welfare interests stem from this fact. However, this is not enough to satisfy Nozick's criterion for inclusion in an analysis of justice, for the importance which he attaches to autonomy is expressly tied in to the idea of the autonomous being's possession of conscious awareness of the meaningfulness of its life and of the connection which the ability to choose its own values has for that meaningfulness. This conscious awareness cannot with any show of plausibility be predicated of non-human animals, not even of those to which a capacity for self-awareness has plausibly been attributed, such as the great apes and pigs. As much may be said of human non-persons, of course, which implies that Nozick's analysis applies to them also.

Thus, the whole of the non-human, and some of the human, are excluded from Nozick's analysis of justice. On his account it is impossible to find room within the scope of justice for those lacking a capacity for autonomy. His argument leads in the direction of the economic market as the crucial social mechanism for the creation and distribution of holdings, and, of course, the non-human world, although made use of within market activity, cannot participate in it in the conscious, autonomy- and rights-recognizing way that alone gives rise to the justice in holdings which eventuates within it.

Nozick, of course, is willing to recognize, and discuss, human moral relationships with the non-human (Nozick 1974: 35–42). But it is quite clear that his whole approach cannot be adapted to encompass the non-human within the scope of justice in the distribution of resources. Of course, there is also much scope for arguing that his theory does not do a defensible job of explaining what justice involves even for the autonomous human beings to which it pays almost exclusive attention, and that this failure derives from an inadequate understanding of the necessary conditions for the exercise of autonomy even in the specific terms in which he characterizes it. However, discussion of that issue will take us too far afield from the matter in hand.

Let us, then, turn to one of his most searching critics, Michael Walzer, whose theory of justice differs sharply from Nozick's at almost every point, and yet, as we will discover, reaches a conception of distributive justice which is no more hospitable than is Nozick's to the concept of ecological justice.

Walzer eschews rights as a fundamental starting point for thinking about distributive justice among human beings. Such rights as appear prime candidates for universality are too few and too thin to sustain any satisfactory theory of justice (Walzer 1983: xv). Although there turns out to be an important place for rights within Walzer's theory, they are what emerges as a conclusion to the theory – they are argued *to* rather than argued *from*. In addition, they are not fixed, but vary from society to society and from time to time within the development of a given society.

Walzer begins in a completely different place from Nozick, and from the Kantian tradition. He begins from human beings considered as social creatures. The lives of such creatures are given their distinctive shape, not from individual choices about values which produce a view of the good life sufficient to confer meaning on the lives of those individuals, but from the joint creation of social meanings (Walzer 1983: 7–8). Human persons are meaning-creators, as Nozick correctly noted, but cannot create meaning except in interaction with each other, which is a thought to which Nozick gives little attention. What meanings human beings cooperatively create in the course of their productive lives will be shaped by their traditions and modifications of those traditions which arise by a variety of means, including, importantly, the practice of immanent critique within their societies.

When human beings, within their societies, create the benefits and burdens whose distribution among them becomes a moral, and specifically justice, issue, the meanings which they have produced in creating those items will, Walzer argues, determine what is held to be the appropriate distribution of the items in question (Walzer 1983: 6–10). Since different items have different meanings for their joint creators there will be no single appropriate criterion of distributive justice across all cases. Each distinct kind of benefit and burden will comprise a distinct 'sphere of justice' (ibid.: 10). Justice is done within each sphere when distributions are in accord with the meaning of the item for its joint creators. Injustice involves the distribution of a benefit or burden within a sphere on the basis of a criterion of distribution appropriate to a different sphere. When this usurpation of the appropriate criteria is done deliberately by powerful groups via their control of specific goods, what Walzer refers to as 'dominant goods', then that group can properly be deemed to be exercising a tyranny within the society (ibid.: 10–20). Tyranny is prevented when goods are not allowed to become dominant, as when society ensures that neither money nor political power nor religious sanctity can enable their possessors to get their hands on all the good things (ibid.: 17–20). For this to take place, groups charged with guarding the criteria of appropriate distribution within each sphere need to be found and empowered. The state has an important role in this, but cannot act as ultimate guardian of the criteria, for it could only do so if given so much power that its own tyranny would be immediately forthcoming (ibid.: 15).

As with the discussion of Nozick given earlier, this summary extracts only a core of Walzer's approach, but sufficient is here available for it to be readily apparent that Walzer's theory also provides no room for a theory of ecological justice. This time the insurmountable barrier is the theory of social meaning. Walzer's theory implies that there can be no issue of distributive justice without the prior creation of social meaning. It is also crystal clear that social meaning can only arise from the interaction of human beings within social situations. Non-human organisms can, of course, easily be the vehicle for such meanings, as when human beings breed certain kinds of plants and animals for specifically human purposes, or create certain kinds of landscapes, such as ornamental parks and gardens. But while human beings clearly interact with the non-human all the time in order to create the items which bear the meanings in question, the role played by the non-human is not that of co-creator of those meanings. That is clearly intended by Walzer to be an exclusively human activity. This is not to say that the non-human may not play some role in unconsciously suggesting to human beings some possible meanings which their creations may bear, as when the creators of aeroplanes are inspired to seek the 'mastery of the air' to be found among the birds. If the non-humans are not co-creators of social meanings with human beings, within human societies, then there can arise no issues of justice between them. Issues of distributive justice can arise between human beings with respect to the non-human, as the latter becomes a vehicle for social meanings, but not actually between human beings and the non-human.

As we noted above with respect to Nozick's analysis of justice, it is possible to take serious issue with the adequacy of Walzer's theory even when considered purely within its own terms. Many problems have been raised about his theory, often focusing on the adequacy of the idea of shared meanings for determining criteria of justice in distribution, particularly within culturally plural societies (Gutman 1995). These objections may all be surmountable, but it does not seem possible for the non-human to be brought directly within the scope of Walzer's theory, given that humans and non-humans do not jointly produce shared social meanings.

Could a Walzerian not reply that there is nothing within the approach advocated by Walzer to prevent a group of human beings creating a sphere of justice within which it was deemed appropriate to attribute claims of justice to the non-human? Indeed, is this not how a Walzerian would be able to make sense of the enterprise in which this book is currently engaged? That is, some human beings will be taken to have created (socially constructed) a view of the non-human in such a way that it makes sense to attribute to the latter claims in justice to certain resources.

But it is hard to make sense of this within the terms of Walzer's theory. A sphere of justice involves the creation of a good with a meaning, and the meaning determines who, among the joint creators of the good, should get different amounts of the good. On that basis, the possibility just introduced would also have to involve the creation of a good with a meaning – but what could it be in this instance? If we say that what has been created is 'the conception of non-humans as subjects of justice', then how is that good to be distributed? The question does not make any obvious sense. The point is that, for Walzer, distributive justice can make sense only among meaning-creators, and that does not include non-meaning-creators to which meanings of various kinds can be attributed. Thus, to bring in non-humans in such a way that they have a claim in justice against human beings with respect to environmental resources, we must find some other way of doing so than on the basis that they share with us in the creation of goods with a certain meaning. This means that we must leave Walzer's ingenious theory behind.

Both Nozick's and Walzer's theories, then, pose insurmountable barriers to the attribution to the non-human of claims in justice against moral agents. Each finds the basis of justice claims to reside in capacities – autonomy in the choice of the good life; the ability to create social meanings – which non-human non-persons cannot possess. They can each find room within the scope of their theories for the attribution of some moral responsibilities of human persons to the non-human, but not of justice. Arguably, however, they have difficulty with the idea that human non-persons can possess claims in justice against the privileged group of human persons with which they are primarily concerned. For example, neonates possess the capacity neither for the autonomous choice of values nor for the social creation of meanings. But, arguably, they do have strong claims in justice for their fair share of resources needed to survive and flourish as best they can (even if they are in a terminally ill condition and thus will never reach maturity as moral beings).

Let us, then, turn to Rawls's theory of justice, and to its development by Brian Barry, to ascertain how far the concept of ecological justice can be accommodated within it. Since Rawls's theory has provided the paradigm of a liberal theory of justice, this discussion will also begin to address the question of how far liberalism can accommodate ecological justice. To anticipate the upshot of the investigation, we will discover that a great deal of liberalism can be retained, but in one crucial area it will have to undergo serious modification. Whether the modification in question is sufficiently drastic to justify the claim that liberalism has in effect been abandoned, and that we must inhabit (to use John Barry's term) a post-liberal world (Barry 1999: 92) in order to achieve ecological justice, is a matter to which we will turn our attention in due course.

7 Liberal theories of justice and the non-human

In many ways the theory of justice which Rawls developed and refined is not obviously any more promising as a basis for developing a theory of ecological justice than are those of Nozick and Walzer. It, too, seems to presuppose that justice is something which fundamentally can obtain only between fully fledged moral persons. As was the case with Nozick and Walzer, Rawls barely discusses the morality of relations between humans and non-humans. Also, Rawls and Nozick share at least one point in common, namely the Kantian starting point that justice concerns the relations between autonomous beings capable of mutual recognition as moral equals. However, it has from time to time appeared possible to find materials within Rawls's approach to facilitate the construction of a theory of justice which allows a natural place for non-persons. To understand these possibilities we need to say a bit more about the development of Rawls's ideas.

As is well known, Rawls's theory went through various stages of development, among which some of the most important occurred after the publication of *A Theory of Justice*. The crucial move here involved the reconceptualization of Rawls's theory by its author as a political rather than a comprehensive, or metaphysical, doctrine. By this he meant that he offered it as a way of understanding how, within pluralist societies, it would be possible to find an 'overlapping consensus' between citizens with different value commitments on fundamental constitutional matters so as to enable a well-ordered society to exist, rather than as a doctrine of the ideal society which could only ever hope to attract the adherence of a limited number of people (Rawls 2001).

Corresponding to these two stages are two ways of looking for room to embody a theory of ecological justice within Rawls's theory. The first is to examine whether his device of the Original Position could allow non-humans entry into the debate on justice, by a suitable thickening of the veil of ignorance behind which the participants in the Original Position are supposed to engage in their deliberations. The second is to analyse the concept of the overlapping consensus within political liberalism in order to find room for the idea that non-humans can have a claim to be treated justly within Rawls's well-ordered society. As it turns out, the former approach has the great advantage, from the point of view of ecological justice, that if it succeeds then non-humans can be given some form of constitutional protection for their interests. However, unfortunately it does not succeed. The latter approach has the great advantage that it can be developed without conceptual difficulty, but the price which has to be paid is that it provides no way to secure any constitutional protection for the claims in justice of the non-human. Let us consider each of these in turn.

The first of these possibilities need not delay us too long. I have pointed out elsewhere the problem with the attempt to thicken the veil of ignorance in the Original Position so that the participants are supposed to be ignorant not just of their age, sex, values, skills and so forth, but also of their species (Baxter 1999: 150–1). This is supposed to allow them to apply the maximin principle so as to determine which is the worst outcome they might face when the veil of ignorance is lifted and to secure the best position for themselves even if that outcome should materialize. Hence they should, one might argue, only agree to something like a principle of equal distribution of environmental resources between species on the basis that the worst outcome would be to be a member of a species with no access to such resources, and which thus would face rapid extinction. One immediate concern which arises when one contemplates this possibility is that there are many difficulties involved in trying to work out what a theory of justice which sought to distribute resources between species ought to look like. But it is a presupposition of this book that these difficulties, formidable though they are, do not amount to a conclusive refutation of the idea of such justice in distribution between species.

However, there is a fundamental problem with this attempted use of Rawlsian devices, namely that the bid to thicken the veil of ignorance in the manner described will not work. The participants in the Original Position must at least be supposed to possess certain specific intellectual capacities, such as the capacity to engage in means–end rationality, to understand various abstractly characterized possibilities, and to operate with the maximin principle. Given these attributions of capacity there is no room for the idea that they could also be ignorant of their species, for they would know for certain that they would be members of species which possess at least these minimal attributes. If it is life on planet earth which they are invited to consider they would know themselves to be at least a member of the genus *Homo*. In response to this, it might be suggested that they might possess these capacities only for the purpose of engaging in the debates of the Original Position, and then possibly lose them when the veil of ignorance is lifted. But this is impossible to make sense of. In what sense would it be the *same* entity that had engaged in the debates of the Original Position and subsequently turned out to be, for example, a bacterium?

In fact there are many serious problems with the Original Position, even if we restrict ourselves to the use which Rawls made of it. As we will see in the next chapter, Barry, following the lead of Thomas Scanlon, takes over the idea of a fundamental constitutional debate without the various ingenious but ultimately unworkable abstractions of the Original Position, and in doing so allows a much more natural place for the interests of non-humans to be located.

What then of the later doctrine of political liberalism and the overlapping consensus? How does ecological justice fare when brought into contact with this later version of Rawls's ideas? Let us consider a recent discussion by Derek Bell which seeks to show that Rawls's conception of political, as opposed to comprehensive, liberalism can give the proponents of ecological justice all they can properly want (Bell 2003). His aim is to vindicate the idea of 'liberal ecologism', which encompasses the two connected claims (1) that individual people can be both liberals and committed to finding within the non-human a 'fundamental locus of value' (ibid.: 1) and (2) that states which are liberal can justifiably pursue policies based on such value commitments without ceasing to be liberal.

The characterization which Bell gives of ecological justice, considered as something which liberal ecologism aims to pursue, is certainly germane to the purposes of this book. On this account the various proponents of ecological justice have argued for the recognition as proper recipients of justice some or all of the following: non-human individuals, species and/or ecosystems. Their aim in doing so has been to secure for such entities the wherewithal to live 'according to their own forms of life' (Bell 2003: 4), and they have argued that the distribution of the goods appropriate to this end should either be egalitarian or in accordance with some principle of moral weighting.

As Bell then notes, Rawls's own, very brief, comments on the nature of human moral relationships with the non-human amount to the idea that we owe at least some non-humans duties of humanity and compassion, but not of 'strict' justice (as Bell notes, what this qualification involves is not entirely clear), for non-humans are devoid of a 'capacity for a sense of justice' (Rawls 1993: 448). This amounts to the admission of at least sentient organisms into the 'community of moral subjects' but excludes them from the 'community of justice' (Bell 2003: 5). As Bell explains, Rawls goes on to say that the justification for this claim takes us beyond the 'contract doctrine' which is essential to his account of justice between human beings, and ultimately involves a metaphysical doctrine 'of our place in nature' (ibid.: 6). This latter is left undeveloped within Rawls's already extensive ruminations, but, Bell suggests, may be intended to leave open the possibility that in some sense the duties which human beings owe to the non-human are indeed duties of justice, albeit of a different kind from the duties of justice which human beings owe to each other (ibid.).

Rawls's later discussion in *Political Liberalism* of the morality of human relations with the non-human makes the issue rather more problematic, at least from the point of view of ecological justice. As Bell explains, Rawls there allows for the reasonableness of a purely instrumental attitude to the non-human, but also allows that other value positions, such as a stewardship theory, can also lay claim to reasonableness (Bell 2003: 7). What emerges in effect is that no one of these positions is incompatible with the requirements of political liberalism, and that the latter allows scope for the adoption of one or other position by particular societies which are established on the basis of political liberalism. As Bell notes (ibid.: 8), for Rawls the issue of human beings' relationship to nature 'is not a constitutional essential or a basic question of justice' (Rawls 1993: 246).

Political liberalism is, in Rawls's formulations, understood not as the embodiment of some universal truth (a 'comprehensive doctrine'), for so understood it can always be reasonably rejected within a pluralistic society in favour of any of the other comprehensive philosophical and religious doctrines available to human beings. Rather it involves the search for basic constitutional principles which all members of such a pluralist society can agree on for the democratically determined conduct of their lives together – an overlapping consensus. Such principles can only be obtained on the basis that the members of such a society are prepared to recognize each other as 'free and equal' (Rawls 2001: 27). Such principles will be free-standing, in the sense that their acceptance does not itself arise from a commitment to a prior comprehensive doctrine. Thus, they represent what all members of such a society can see to be a fair basis upon which to organize their lives together. Such a politically liberal society is precluded, therefore, from embodying within its constitutional arrangements any principle which is only acceptable if you are committed to a specific comprehensive view.

On this understanding of political liberalism, the appropriate place for specific comprehensive doctrines to exert their influence is within the procedures established for democratically deciding upon specific policies. As Bell notes, this means that the earlier 'metaphysical view' within which Rawls anticipated that it might be possible to find a defensible doctrine of human moral relations with the non-human, and which he envisaged as an addition to his theory of justice, is now definitively placed within the realm of reasonably contestable comprehensive doctrines, none of which should be allowed to influence the basic design of pluralistic democratic societies.

Bell in effect accepts the cogency of Rawls's reasoning here, and proceeds to introduce the idea of ecological justice as precisely one of those comprehensive doctrines which are entitled to have their opportunity to influence, and perhaps determine, the course of policy-making and legislation within a pluralist society. As he says: 'Liberal citizens are entitled to vote their ecocentric ideals and a liberal state may promote ecocentric policies that have been endorsed through the democratic process' (Bell 2003: 10).

Bell notes that there might appear to be a problem lurking in this suggestion, in that political liberalism might appear to be committed to the view that the only proper place for the concept of justice to operate is within the concept of political justice, and thus the notion of ecological justice, if this is precluded from operating to determine the fundamental political structure of the liberal society, will have to be rejected. His solution is in effect to be relaxed about the use by political ecologists, or by proponents of any other comprehensive doctrine, of the term 'justice' within the articulation of those doctrines. Provided that all supporters of specific comprehensive doctrines accept the political conception of justice, and thus are prepared to work within the framework of the politically liberal state, they can use whatever terms seem suitable to them to express the moral dimensions of their doctrines (Bell 2003: 10–11).

Bell then notes some further implications of this. The first is that liberal ecologists must give priority to political justice (establishing the fundamental constitutional arrangements of their society) over ecological justice, and since the former is exclusively anthropocentric (its concern is solely with the political conditions essential for human well-being within plural societies) this means prioritizing human interests over those of non-humans. This, however, Bell argues, is only a problem for liberal ecologists who are moral egalitarians with respect to the non-human world. Those who reject such egalitarianism typically give priority to non-humans in any case (Bell 2003: 11). Nor is it necessary to suppose that Rawls's version of political liberalism, containing as it does the Difference Principle, commits political liberal societies to securing for their citizens as high a standard of living as possible, so that the interests of non-humans must inevitably be crowded out. There is nothing, in other words, in the very idea of a politically liberal society which would preclude human beings' collectively choosing to forego certain possibilities of material advancement in order to secure the availability of scarce resources for the benefit of non-humans. Hence, Bell concludes, there is nothing in the pro-human biases of political liberalism which should prevent political ecologists from accepting it (ibid.: 12–13).

Finally, Bell seeks to vindicate the refusal of political liberalism to admit the non-human into the deliberative procedures of liberal societies via human proxies of one kind or another. He claims that the starting point of political liberalism has to be the interests of human citizens of a pluralist society who are trying to discover some principled means of living in a peaceful and orderly way with each other in spite of their diverse and sometimes conflicting value commitments. Thus he firmly rejects attempts to argue that the interests of non-humans should also be introduced at the fundamental level of constitutional debate.

Interestingly, however, Bell does not do so on the basis one might anticipate from the general logic of political liberalism that he has already presented, namely that any attempt to do so could only be justified on the basis of the prior acceptance of some comprehensive view which not all could be expected to accept. Rather, he argues that the political representation of the non-human would be unjustified, and he presumably believes that this argument ought to convince anyone who attends to it irrespective of their specific value commitments. The reason it would be unjustified is that, while both moral agents and those beings which are only moral patients (he uses the term 'moral subjects' instead) can have interests which can be harmed, and while both are morally considerable, only moral agents can be harmed by exclusion from participation in the decisions which affect them. As he puts it: 'If they are excluded from the moral community, their capacity (*qua* moral agents) to take responsibility for their actions is not respected' (Bell 2003: 14). This form of injustice cannot be perpetrated against moral patients which cannot take responsibility for their actions. Non-humans are entitled to some protection of their interests, but only human moral agents are entitled to constitutionally protected representation of their interests.

The obvious concern here for proponents of ecological justice is that if there is no built-in representation of non-human interests then there is nothing to ensure that the interests of the non-human will receive significant protection by policy-makers and legislators. However, Bell argues that this, while true, can only be remedied at the expense of democracy, by ensuring that those human beings with a commitment to ecological justice are placed in unassailable positions of dominance within the procedures for making policy and legislative decisions.

Since most liberal ecologists will see that there are serious problems with attempting this route they will in practice fall back on the only avenue open to them within political liberalism, which is to seek to persuade their fellow citizens as best they can within the normal procedures of democratic political debate to protect the interests of the non-human.

This argument of Bell's does serve to clarify many of the issues with which a proponent of ecological justice has to deal. Some preliminary responses may be made before a more thorough examination of these issues is undertaken in the next chapter in the course of examining Barry's version of political liberalism.

The first point is that Bell's argument against the very idea of representing, as opposed to simply protecting, the interests of moral patients is open to objection. It is true that injustice is done to moral agents when they are prevented from being represented in decision-making forums where their interests may be affected. But that is not because, as Bell puts it, 'the purpose of inclusion is to recognize a moral agent's responsibility for her conduct' (Bell 2003: 15). The purpose of inclusion is rather that she can articulate her interests and no other moral agent can claim superiority to her in this respect. Failure to include her treats her as if she did not possess capacity for autonomy, which is to fail to treat her with the respect which is her due. It is true that we do also hold beings with the capacity for autonomy to be responsible for their actions, at least most of the time, but that is not the reason why it is appropriate to grant them the right to participate in making the decisions which affect them. If this is right, however, it means that the focus in arguing for representation is on interests, not on responsibility. This is as it should be, for it can plausibly be argued that human beings who do not possess autonomy, such as babies and very young children, also need to have their interests represented in decision-making forums within a politically liberal society, inevitably by human proxies. Arguably, too, it is appropriate to seek to find a means of constitutionally protecting the right of non-autonomous human beings to the representation of their interests within the decision-making procedures of the politically liberal society. Their inability to articulate their interests for themselves should not preclude the representation of those interests in the decision-making forums where those interests may be directly affected.

This leads to the second point of objection to Bell's argument, namely his rapid move from the idea that exponents of ecological justice may be concerned to ensure that the interests of the non-human are taken into account to the conclusion that this commits them to some form of authoritarian or oligarchical rule at odds with the fundamentally democratic structure of politically liberal society. This is too rapid a move, because limitations put upon the workings of democracy in order to prevent democratic majorities from ignoring the rights of certain members of the society are not properly thought of as anti-democratic in any sense. Such constitutionally embedded rules are better seen as obstacles to the tyranny of the majority. Within the terms of political liberalism such constitutional embedding would be objectionable if it could be shown that the rights being defended are acceptable only if one has a prior commitment to some comprehensive view, for basic constitutional structures are not supposed to involve commitments to such views. However, as we noted above, at least in Bell's argument that is not the point of objection.

Thus, if the point of representation within decision-making forums is to articulate the interests of all those – human and non-human, autonomous and non-autonomous – whose interests are affected by the decisions to be reached within those forums, and if it is not automatically anti-democratic to ensure by constitutional provision that the interests of the non-autonomous, whether human or non-human, are taken full account of within the democratic institutions of a politically liberal society, then it is not objectionable for exponents of ecological justice to do precisely that.

Of course, Bell is quite correct in saying that the only permissible means to attain such an outcome where it does not already exist within a basically politically liberal society is for liberal ecologists first to persuade their fellow citizens to agree to the appropriate constitutional amendments. This will, of course, be difficult – but not impossible, as the example of the German constitutional amendment to secure rights for animals shows (Connolly 2002). And if it is achieved then that does not guarantee its own permanence, for constitutional provisions can always be overthrown. But this does not mean that liberal ecologists who pursue this route are engaging in some futile endeavour, for constitutional amendments do make certain kinds of morally objectionable behaviour more difficult. They can also have this effect even when they are initiated by officers of the liberal society who do so as a professional duty, rather than as a matter of personal conviction, so that we do not need to suppose the necessity for a policing body made up of zealots. Nor does it mean that liberal ecologists can properly be branded as authoritarian, any more than the authors of bills of rights in the purely human case can properly be so characterized.

However, the claim may still be made that the conception of the overlapping consensus which underlies political liberalism is sufficient to rule out this version of liberal ecologism. For it might well be reasoned that the arguments in favour of the constitutional protection of rights to representation for non-human interests do indeed contain a commitment to a comprehensive view which it would be unreasonable for those who believe it to expect all of their fellow citizens to accept. In order to evaluate this argument, however, we will do well to consider the version of political liberalism to be encountered in the work of Brian Barry, for that shows an awareness of the difficulties for liberalism that are posed by the existence of non-autonomous humans which, when taken proper account of, show that the objection just given cannot be sound.

However, as we will discover, there is a sting in the tail for political liberalism when we do take this into account. This is that there are serious problems with the crucial idea of the overlapping consensus which do not derive from the existence of competing comprehensive views, at least when these are conceived of, as they usually are, as articulating a view of the 'good life'. There is the possibility of ineradicable differences of opinion about procedures, especially about whose interests are counted in political debate, which make the eirenic purposes of Rawls's procedural liberalism unattainable. This means, in turn, that, although proponents of ecological justice can and should accept a large amount of the standard ingredients of the politically liberal society, in the end they find themselves having to take a stand upon matters of substantive moral disagreement with many of their fellow political liberals about the precise content of political liberalism. Essential contestability turns out ultimately to be unavoidable.

Before we turn to a consideration of Barry's argument, however, it will be useful to give some thought to an issue which has been lurking beneath the surface of the discussion thus far. This is the question of how far liberalism is actually committed to the values which are embodied in the concept of ecological justice. Bell, Barry and Rawls all take the view that a commitment to the moral considerability of nature in general and of any sub-part of it in particular is not ruled out by political liberalism. Political ecologists are allowed their opportunity to put forward their case to their fellow citizens and seek to protect what they take to be under threat by means of the normal forms of legislation within a liberal society. But there are two other possibilities with respect to the connections between liberalism and the non-instrumental value of nature which need to be considered besides this permissive one. The first is that a commitment to liberalism is incompatible with any very extensive moral valuation of the non-human. As Bell suggests (Bell 2003: 1), most versions of 'green liberalism' (see, for example, Wissenburg 1998) are of this sort. The second is the opposite view – that a commitment to liberalism actually requires an acceptance of the moral considerability of nature, or at least some aspects thereof. Let us consider the case which can be made for this claim.

The view that the valuation of nature 'for itself' is in some way appropriate for a liberal to believe, a liberal in the political rather than the comprehensive sense, even if such a belief is not exactly a logical implication of liberalism, has recently been championed by Simon Hailwood (Hailwood 2003). As he explains, he takes Rawls's political liberalism as the 'paradigm liberal theory' and assumes that such liberalism can be a theory of environmental justice (ibid.: 1). He assumes, further, that ecological justice involves 'a non-instrumental view of nature (Nature does not have the status as a mere resource for humanity; doing justice to the non-human world involves recognizing this, at least as a first step)' (ibid.). He then aims to show that 'these two kinds of justice can be viewed as continuous with each other' (ibid.). This does not set up the problem in quite the same way as this book does, where the focus has been on the moral considerability and claims in justice of non-human organisms, rather than of 'nature'. But this appears not to affect the course of the argument very much, as we will discover.

The next key stage in the argument is to defend the non-instrumental view of nature to political liberals. To this end, Hailwood invokes the 'otherness view' of nature: 'nature's very independence of humanity – its otherness – is a ground of non-instrumental value' (Hailwood 2003: 1). Hailwood's use of the term 'nature' here is intentional. He does not intend to refer simply to organisms, but also to abiotic nature. As he expresses the point: 'abiotic nature has no "interests" to take into account . . . still, respecting *its* otherness involves an objective recognition that it is not merely a human resource' (Hailwood 2003: 22, n. 45). It is not part of the argument of this book that abiotic nature is an appropriate subject of ecological justice. A discussion of how appropriately to view it would take us too far afield at this point. Suffice it to say that Hailwood's claim can obviously be reformulated in the more specific way germane to the argument of this book as follows: non-human organisms' very independence of humanity – their otherness – is a ground of (their) non-instrumental value. Why should a political liberal accept this claim?

Clearly, it is not a matter of serious dispute that other organisms are independent of humanity. If human beings were to disappear from the planet forthwith, most organisms (excluding our commensals and parasites, perhaps) would survive, just as the world contained countless organisms for billions of years before *Homo sapiens* emerged. We all, including political liberals, have to recognize this fact – but why do political liberals have to respect this otherness, rather than just recognize it? Why, to cite Hailwood's most ambitious claim, is it arbitrary for political liberalism to 'exclude respect for nature's otherness (as something like a constitutional commitment)' (Hailwood 2003: 2)?

The answer which Hailwood gives to this question is to point to what he sees as a 'congruence' between the attitude of respect for nature's otherness and two connected commitments essential to political liberalism's conception of reasonableness. These are anti-expressivism – one should not expect to identify with the state on the basis that it expresses one's specific value commitments – and the neutrality of the state between competing comprehensive views of the good life (Hailwood 2003: 1). Certainly these commitments are definitive of political liberalism, but what exactly is the force of the 'congruence' which Hailwood detects between these commitments and the attitude of respect for the otherness of nature? Congruence is literally a matter of a similarity in form between two different items. Is this sufficient to tie together value commitments embodied in each of the items between which congruence is detected?

The form of this argument is an onus-changing one. What is happening is that the onus is being placed upon those who believe that political liberalism can allow nature only to have instrumental value (let us call them 'instrumental political liberals' for short) to prove their case. It does not appear, on Hailwood's argument, that there is any formal inconsistency in the theoretical position of instrumentalist political liberalism. But if Hailwood is correct then non-instrumental political liberals are not guilty of any inconsistency either. Moreover they have an advantage over the instrumentalist in that they can produce the congruence argument – respect for the otherness of nature, involving recognition of its neutrality between different human purposes (Hailwood 2003: 7), is akin to the neutrality of the state between different substantive views of the good life.

The neutrality of nature is explained as follows. Nature has been made over in a variety of ways by human beings, but it is always possible to disentangle nature as it objectively exists in independence of any particular human purpose and the specific purposes to which human beings may put it. The same natural substrate can be put to a variety of human purposes (although there may be some limits to uses which arise from the characteristics of this substrate), and it is thus properly to be thought of as neutral between any of them (Hailwood 2003: 7). One of the implications of this is that human beings cannot read off any moral lessons from nature, in particular that human societies are better the 'nearer' they are to nature (ibid.: 8).

However, although instrumentalist political liberals are not guilty of internal doctrinal inconsistency, Hailwood does want to convict them of the kind of arbitrariness which is characteristic of human chauvinism: 'either an unwilling-ness to acknowledge the existence of nature independent of landscape altogether, or an assumption that nothing matters unless it plays some role in human landscapes' (Hailwood 2003: 6). Arbitrariness in regard to an intellectual position means that there is no good reason for holding it. Hailwood recognizes that this argument cannot force instrumental political liberals to abandon their position on pain of intellectual incoherence (it rather, he says, 'chips away' at the instru-mentalist position as the 'automatic default position' for political liberals (ibid.: 6, n. 12). But one may well wonder which of these arguments is going to do the real work of convincing instrumental political liberals to go over to the non-instrumental position – the positive 'congruence' argument, or the negative 'arbitrariness' argument.

The latter as Hailwood characterizes it does not seem quite up to the job in any case. What is rather needed are those kinds of argument which have been championed by many environmental ethicists who urge the moral considerability of non-human nature. The earlier chapters of this book urged this also in the company of some excellent theorists, such as DeGrazia, Warren, Wetlesen and Singer. If their arguments are sound, then human chauvinism is worse than arbitrary. For it involves the rejection of strong positive reasons for recognizing the moral considerability of non-human organisms.

An important point about these arguments in the present context is that they do not make any play with the 'otherness' of nature as such – its independence of human purposes and plans. Such otherness, although undeniable, does not in itself justify any positive valuation of non-human nature. Rather, 'otherness' in this sense is an indication of the more fundamental idea of moral considerability. Thus what does do the job of enjoining the positive valuation of non-human nature are the arguments for recognizing (1) the existence of welfare interests among all non-human organisms; (2) the ability of human moral agents to harm or benefit those interests; and for (3) that these two facts give a positive reason for human moral agents to regulate their behaviour so as not to harm those welfare interests without good moral cause shown. Indeed, in the deployment of the casuistic approach of Wetlesen, it turned out to be crucial to discover similarities between human persons and non-human organisms to motivate the positive valuation of them in moral terms. The existence of such similarities is not, however, incompatible with the 'otherness of nature' thesis, which requires only that non-human nature has purposes independent of those that characterize human beings.[1]

Hailwood moves towards these positions at various points in his discussion without bringing out their full import. For example, he tells us that

> To say nature is non-instrumentally valuable in virtue of its otherness is to claim it is an 'end in itself' . . . I am taking it that nature is a negative end in virtue of its otherness: to the extent that it is treated as such an end it is 'not

interfered with' in the sense of ridden roughshod over or altered without constraint (and this for its own sake and not only because it might be imprudent for us to do such things).

(Hailwood 2003: 7)

This is the kind of claim that underpins the argument for ecological justice, but it needs to be more fully explained in terms of more specific categories than can be encompassed in the blanket notion of 'nature'. Here is where it starts to become imperative to disaggregate the 'end in itself' of 'nature' into the several ends in themselves of specific components of the non-human world.

It is perhaps also important at this juncture to note that even an instrumentalist political liberal can find instrumental reasons for valuing the otherness of nature, a point to which Hailwood himself draws attention in connection with Goodin's views expressed in the latter's *Green Political Theory* (Hailwood 2003: 17; Goodin 1992: 39). Human beings do undoubtedly find psychological or perhaps spiritual refreshment in encountering other creatures which are pursuing their own purposes in complete indifference to human interests. Such encounters can enable us to set aside our own worries and problems and in imagination see the world from a completely different perspective, perhaps returning to our own concerns with a more balanced view of their character. Valuing nature's otherness in this way might well lead one to seek to preserve the possibilities for its own independent existence, and thus might take one in the same direction as ecological justice concerns. But it does appear, as with all instrumental arguments, to be too contingent to satisfy the concerns of ecological justice. After all, not everyone finds it spiritually refreshing to experience the otherness of nature – it frightens, alienates, depresses or annoys at least some people. Thus, the 'arbitrariness' argument deployed by Hailwood, while the right kind of argument to take instrumental political liberals in the direction of ecological justice, needs to be considerably strengthened to do the job required.

What then, to return to Hailwood's more distinctive contribution, of the 'congruence' argument, which after all is fashioned precisely to target political liberals, not human chauvinists in general? In one way it is possible to exhibit a pleasing symmetry between the political structure known as the state, as understood in political liberalism, and nature as explicated in Hailwood's analysis. Both are, one might say, necessary conditions for the actualization of the liberal polity. The state facilitates by means of a regime of reasonable basic rules the pursuit by individuals of their views of the good life, and remains neutral between them; nature facilitates by means of the provision of material substrates the same pursuit of different views of the good life, and remains neutral between them. A key difference remains, of course, between the neutral state and neutral nature, namely that, as Hailwood has been at pains to emphasize, nature has its own existence independent of human purposes, while the state exists only because human beings create and sustain it. For the neutral liberal state to acquire its own independent purposes would be a sure sign, from the political liberal point of view, that something had begun to go badly wrong with the political system.

This alerts us to a possible rebuttal which the instrumentalist liberal may be tempted to offer to Hailwood's position. Granted that nature is a necessary and neutral material substrate of the liberal polity, the fact of its underlying otherness, a feature which does not apply to the state in a properly constituted liberal society, should be seen for what it is – the ever-present possibility of an alien and hostile force operating within human life. The very fact that nature is 'other', and is never in fact, or even in principle, fully made over in the light of human purposes, is why we should never allow ourselves to become trammelled by an untenable moral restriction on how we interact with it. We have to make use of nature – that is agreed on all hands. But nature opposes human purposes (not of course wilfully) as much as it facilitates them. For this reason it is entirely reasonable to take the view that the non-human components of nature, each with their own, nearly always blind, purposes have to look after themselves. That is the flip-side of their 'otherness'. Nature is not a moral agent, capable of restraining itself in the light of reasonable principles which it could come to agree upon in the course of discussion with human beings. That too is agreed on all hands. Hence the relation between human beings and nature is always going to have to be one of uneasy coexistence, unmediated by any real moral connection. We should take care of nature, but only for our own prudential reasons, not for any moral duty we may be held to owe it. Thus, the political liberal has conclusive reason to recognize the otherness of nature, but no good reason to respect it, morally speaking.

How might a defender of Hailwood's position respond to this attempt to counter the 'congruence' argument? First of all, he has available one possible parallel between human–nature relations and human–human relations within the polity of political liberalism, namely that in the latter one implication of allowing others to pursue a view of the good life with which we do not ourselves agree is that we have to put up with activities which we find offensive or shocking. We can find each other to a degree to be hostile and alien, but recognize the importance of mutual toleration to the possibility of a well-ordered society within which we too can pursue our view of the good life. So too with our attitude to nature's otherness, where a form of toleration is also required:

> Here . . . it does not seem out of place to talk of *tolerating* nature as other; as a matter of attempting *not* to stamp out its every inconvenient encroachment, obliterate every annoying natural obstacle, or recoil . . . from every encounter with the non-human in its more apparently aesthetically and morally sub-optimal aspects.
>
> (Hailwood 2003: 14; original emphasis)

Hailwood also has available a version of the 'integrity' argument deployed by Tim Hayward. The reasonableness of political liberalism involves the development and embedding within the human psyche of the distinctive attitude of toleration, which has an extremely demanding and complex character. It is not just a matter of putting up with what one experiences as the offensive, shocking and inconvenient behaviour of others. It is also a matter of dealing appropriately with the behaviour of fellow citizens which steps over the boundary of reasonableness (as when they mount an attack on the right of their fellows to pursue their own view of the good life) without allowing the existence of such behaviour to poison one's general attitude to one's fellow citizens, to the extent of seeking authoritarian or draconian responses to such problems. This, one can say, is a matter of not allowing oneself to become corrupted by the experience of the less-than-optimal behaviour of one's fellows. It is, in other words, a matter of maintaining one's moral integrity.

Thus, in the purely human case, the persons of moral integrity maintain an appropriate firmness in dealing with unreasonable (in the political liberal sense) behaviour, while maintaining their respect for alternative views of the good life which they find personally distasteful. Hailwood makes this point in terms of the Rawlsian notion of the 'political capital' of society (Hailwood: 14). The political liberal is a person with certain specific virtues which are necessary to the continuation of the liberal society, and the widespread existence of such virtues is 'a great public good' (Rawls 2001: 118). Once this point is granted, then it follows unproblematically that it is appropriate to look for the same integrity, the same virtue, in the response to non-human nature of human beings who are members of the political liberal society. One will respect nature's otherness, tolerate its inconveniences and occasionally shocking aspects, and deal appropriately with its unthinking attacks on the interests of human beings without allowing oneself to be dragged into an authoritarian or uncaring attitude towards it. As Hailwood puts the point:

> Seen like this then, respect for (involving toleration of) otherness looks continuous between the realms of people, state and external nature. None of these are just there to be used for the purposes of particular individuals and groups . . . even though it might often seem easier, instrumentally rational, to do so.
>
> (Hailwood 2003: 14)

However, if this is the appropriate argument to deploy against the political liberal's rebuttal to the congruence claim, then it looks as if it is rather stronger than Hailwood's deployment of the concept of 'congruence' would indicate. For it is not that, in a liberal polity, the virtue of toleration towards nature is congruent with the virtue of toleration towards fellow citizens, it is that it is the very same virtue! The point is brought out most fully if we employ Tim Hayward's concept of 'integrity'. For it is then readily apparent that what is involved is the unity of moral individuals as they exercise a virtue with respect to different but equally appropriate objects.

It might be thought that the fact that nature differs from one's fellow citizens, in not being a moral agent one can hold accountable, remains an important distinction sufficient to ground a difference between our attitudes to nature and our attitudes to our fellow citizens. But this is not so, for many of our fellow citizens in a political liberal polity are not moral agents either, temporarily or permanently, and may do us harm, intentionally or otherwise, and yet they are still fellow citizens. The importance of such human non-moral agents to the polity of political liberalism is something we will explore more fully in the next chapter.

The main thrust of Hailwood's argument, then, is an important and sound one. The political liberal has sufficient reason, from within the doctrine of political liberalism itself, for respecting and not simply recognizing the otherness of nature. This means recognizing nature as possessing non-instrumental value, as an 'end in itself'. This then leads Hailwood to conclude that, within the state constructed by political liberals, 'respect for nature's otherness . . . should be something like a "constitutional essential" alongside the standard commitments to freedom and equality within liberal theory' (Hailwood 2003: 22–3), and he notes that an argument of the kind he has made is applicable to kinds of liberalism other than the political conception upon which we have been focusing in this chapter – and will continue to focus on in the next (Hailwood: 23, n. 46).

These are all points with which defenders of ecological justice should agree. The practical implications of the position are indicated by Hailwood as being 'limiting the destructiveness of human activity' (Hailwood 2003: 19) in various ways. But he is chary of trying to work out any detailed principles of ecological justice, which is a task best left to 'democratic deliberation' (ibid.: 23). We will not be so cautious, and an attempt will be made shortly to bring out some general principles of ecological justice as applicable to different kinds of organism.

Conclusion

In this chapter we have seen how the version of Rawls's theory of justice which viewed it as an argument for political, not comprehensive, liberalism appears to allow various possibilities for the acceptance of ecological justice within a liberal society. We saw reason to reject Bell's argument that the appropriate place for it is within the realm of competing conceptions of the good. Rather, proponents of ecological justice need to find a way of locating it within the fundamental constitutional provisions of a political liberal society. We examined, in the light of Hailwood's discussion, the case for saying that the theoretical commitments of the supporter of political liberalism also commit such a supporter to respecting the purposes of nature which are independent of human beings. This case, as we have seen, upholds the claim that constitutional recognition of the need to respect nature should be part of the political liberal society.

However, there are likely to remain serious qualms with such a conclusion among many who support the Rawlsian case for political liberalism. The above attempt to use the 'integrity' version of Hailwood's argument might be rejected, and the sole version accepted (if any is) might be that which sees only a 'congruence' between the otherness of nature and liberal political reasonableness. This would allow supporters of instrumental political liberalism to claim that they have not yet been convicted of any intellectual fault in adopting their position. They would remain at liberty to support the neutrality of the state with respect to competing views of the good life while rejecting the claim that they should respect the otherness of nature.

A further objection is likely to stem from the claim that the conception of reasonableness which lies at the heart of the Rawlsian conception precludes the formulation of constitutional principles which pay any attention to anything other than what may directly affect beings capable of such reasonableness, a restriction which excludes any element, biotic or abiotic, to be found in non-human nature.

With respect to the former, there does not appear to be much more to say. If the argument is accepted that the virtue which is to be exercised within political liberalism is not the same as the virtue which is to be exercised in the respect for the otherness of nature, then we appear to have something of an impasse between the two positions. Let us, then, try to deal with the second kind of objection. If we can show that the reasonableness condition, which is central to the contractual situation at the heart of political liberalism, can encompass beings themselves incapable of exercising such reasonableness, then we may have found a way of requiring political liberalism to recognize the case for ecological justice within its constitutional provisions. To examine these matters let us turn next to consider what is arguably the most carefully worked-out version of political liberalism – that to be found in the work of Brian Barry.

8 Ecological justice and justice as impartiality

The aim of this chapter is to try to relate the idea of ecological justice to what is arguably the most defensible theory of distributive justice developed for the purely human domain in recent years. This is the theory of Brian Barry, which he has dubbed 'justice as impartiality' (Barry 1995). It is a theory of justice which takes much of its inspiration from Rawls' theory of justice as fairness, albeit via some important modifications of that theory introduced by the political philosopher Thomas Scanlon (see ibid.: 67–72).

There will not be room here to develop the case for accepting Barry's theory against its competitors, but the issue of its merits will be briefly returned to after an outline has first been given of its main features. The guiding thought behind what is here being attempted is that if the concept of ecological justice should prove to be wholly incompatible with Barry's theory then defenders of the concept will have an even bigger job on their hands than they might otherwise suppose. For they will have to develop their own theory of distributive justice for human beings which is defensible for that domain and at least compatible with, and preferably supportive of, ecological justice. Given the enormous amount of intellectual effort in the last thirty years which has gone into developing the ideas that have emerged in Barry's theory, it is reasonable to suggest that it would be preferable if defenders of ecological justice could marry this theory to their concerns, rather than have to start again from scratch.

As will emerge in due course, this marrying is possible. But there is a significant cost for Barry's theory in seeking it. For it is thereby deprived of the possibility of realizing one of its key aims, namely to stand in a position of lofty detachment from all the substantive moral views which divide people and show how such people can, nevertheless, be reconciled, even when their moral views remain irreconcilable.

Barry's theory of justice as impartiality

Let us first, then, consider the basic elements of Barry's ambitious theory. We are invited to imagine, in accordance with the idea of the social contract tradition as updated by Rawls, that a group of people want to live together on terms which all can accept as reasonable. They will come to see that a fundamentally just society will be one which all can freely agree to live in because the basic rules under which it operates treats everyone as moral equals. Hence, no one will be able intelligibly to claim that their interests are being unfairly ignored or suppressed by the basic structure of their society. As Barry expresses the point: 'The essential idea [of justice as impartiality] is that fair terms of agreement are those that can reasonably be accepted by people who are free and equal' (Barry 1995: 67–72).

The chief difficulty in devising reasonable rules is, as Rawls saw, that people hold different and often conflicting views of what values are the important ones and of the best way to ensure human flourishing – in a word, of the good. Hence, they will tend to support only those rules which are at least compatible with, and preferably support, their own view of the good.

As we noted in the last chapter, given this problem, Rawls devised the concept of the 'original position' in which he stipulated the existence of a 'veil of ignorance' which prevented the people engaged in constitutional debate from knowing their own view of the good. This device was intended to embody the basic requirement of fairness – no substantive view of the good is given precedence over any others. Barry has many criticisms of these Rawlsian devices which we do not need to go into (see, for example, Barry 1995: ch. 3). However, he accepts the basic idea of Rawls's theory, specifically the idea that just rules for a society are those which are fairly arrived at. Their fairness is determined by the fact that no one can be supposed to have reasonable grounds to reject them, even when they in fact are likely to fare least well under them in the actual society which emerges. As Barry puts it: 'I shall use the negative formulation [of non-rejectability] whenever I want to emphasize the point that in justifying a principle special attention must be given to the problem of justifying it to those who stand to do the least well out of its operation' (ibid.: 70).

Any proposed rules, then, are to pass the test of being 'not reasonably rejectable and thus fair'. In Scanlon's theory Barry finds a version of the social contract idea which avoids the criticisms that he levels at Rawls's theory. In this version, the would-be members of the society do know their own views of the good and are equally well informed about the likely effects of various possible rules on their well-being and interests seen from the perspective of their view of the good (Barry 1995: 67–72). They are each supposed to have a veto over the proposed rules, and to be motivated by an important consideration which Barry labels 'the agreement motive', that is they wish to live peaceably together under a constitutional arrangement which they all freely accept (ibid.: 164–8).

Barry, pursuing the Rawlsian insight, then reaffirms that the only way in which this desired outcome can be achieved is if the people in question devise rules which do not favour any substantive view of the good. This in turn entails that the fairness of the rules arrived at must be logically separate from any such substantive view. Their acceptance, then, is based on the view that they are reasonable principles for all to live under. To put the point the other way round, it means that they are rules which no one can reject, or veto, as being unreasonable.

Albert Weale has usefully outlined how Barry understands the idea of what people can reasonably object to (Weale 1998: 21–2). All the Scanlonian contractors, as we may call the parties to Barry's version of the social contract tradition, will agree that it is reasonable for contractors to veto:

- something which threatens their vital interests;
- something which unfairly privileges views of the good other than theirs;
- something which prevents the provision of public goods or does nothing to prevent free-riding on such provision.

Weale further suggests, with some plausibility, that, for Barry, reasonableness in the Scanlonian contract situation involves the conjunction of two principles:

> Reasonableness therefore means that parties in the original position do not think of themselves as infallible, and in making proposals for the basic structure of social and political organization they seek to act in a way in which they accept the burden of justifying their own point of view to others.
>
> (Weale 1998: 25)

Given that, according to Barry's theory, the basic constitutional rules and institutions under which the people are to live can be shown not to favour any substantive theory of the good over any other, it follows that the theory of justice as impartiality itself does not presuppose any theory of the good. It is, Barry claims, 'free-standing' with respect to the good (Barry 1995: 76).

This means in turn that the theory of justice as impartiality is not, and is not meant to be, the whole of ethics (Barry 1995: 77). Its aim is simply to establish a framework – the only possible one, in fact – within which people can freely and rationally determine (in the form of policies, laws and permissions) their joint answers to the other issues of ethics, in the light of their free political debates on the rival merits of different theories of the good. Hence, although the theory is limited in scope, what is claimed for it within those limits is extremely ambitious.

What basic constitutional rules are supposed to emerge from the fair debate among people subject to the agreement motive, desiring reasonable grounds of association, and each possessing a veto power over constitutional proposals? Unsurprisingly, given the origins of the theory in the tradition of debate within liberalism about the requirements of justice, the constitution which emerges is recognizably (and is expressly intended by Barry to be) a liberal one.

Thus, certain constitutional provisions will forbid substantive views concerning religious belief and sexual practices from having any effect on law and policy (no banning of religions; no banning of any form of consensual sex) (Barry 1995: 82, 84). This is because of the centrality to people's lives of their being able to choose and/or practise their preferred religious faith and sexual orientation. However, in most other areas of public policy (including protection of the environment and other species) Barry's theory has no such implications. Instead, it requires constitutional provisions and other institutional arrangements (governing, for example, the education system and the mass media) to ensure that certain procedures are followed for the formulation of law and policy in all these areas. These procedures involve, roughly, free, well-informed and fair debate within a democratic framework. In other words, justice as impartiality underpins a concept of procedural justice for the determination of policy in such areas (ibid.: 110).

As we saw when we considered Bell's characterization of political liberalism in the last chapter, within the resulting society devotees of the various substantive theories of the good will then be able to do what they legitimately can to influence their fellows in their deliberations. But there is no guarantee of success for any substantive theory. However, as long as the procedures mandated by Barry's theory are followed, and the constitutional provisions respected, any particular outcome, although morally bad from the point of view of specific theories of the good, will be in accord with the requirements of justice. As Barry puts the point, any particular outcomes will probably be regarded by some of the citizens of the just society as bad, but they will also accept them as legitimate (Barry 1995: 150).

Barry recognizes that there undoubtedly are alternative views of justice whose supporters will necessarily regard such outcomes as not just bad, but specifically unjust. He explains this possibility as depending upon a commonly held view of justice, different from his own, which subordinates it to the good (Barry 1995: 76). Thus, to cite his example, an unreconstructed Thomist will see justice as requiring the acceptance and implementation of Thomist views of the good. Such people will prima facie have no good reason to accept justice as impartiality precisely because it does not guarantee the implementation of Thomist prescriptions.

However, Barry suggests, even such people may come to recognize the force of justice as impartiality, as soon as they realize that:

- there is no realistic prospect of their gaining universal voluntary acceptance of Thomism;
- that there is no prospect of forcing it upon people; and
- that they may themselves face persecution for their beliefs in a society which offers no protection to people's holding whatever views of the good seem fit to them.

These considerations provide the prospect of their coming to have the motive already mentioned whose force is essential to Barry's theory – the 'agreement motive': the desire to live peacefully with fellow citizens who do not share their views of the good (Barry 1995: 122–3; 162–3).

Barry has at various points in his recent writings used examples from environmental matters in the real world to illustrate further the implications of his theory (see Barry 1995: 145–51; Barry 1998: 241–3). We have already encountered an account of them in Bell's analysis, and they do not make entirely comfortable reading for those who are concerned with ecological justice. To put Barry's view on issues concerning the preservation of endangered species, habitats and ecosystems in a nutshell, he says that whether or not any component of non-human nature is saved depends on whether or not enough people can be persuaded of what he calls the 'ecocentric theory of the Good'. On his interpretation of it, this is the view, to which he says he is himself attracted, which accords the interests of non-human species moral weight. If enough humans can be persuaded to accept this, he claims, non-human nature will be saved; if not, not (Barry 1998: 243).

Of course, this is a truism. But the key point is that, according to Barry, either outcome is compatible with justice as impartiality on his account of it. That is, justice in this sense may be done whether or not human beings exterminate large numbers of other species. Further, if insufficient numbers of people are convinced of ecocentrism, then the concept of justice places no real obstacle in the way of such exterminations. Many environmental ethicists will feel that something has gone wrong here, precisely because they will be strongly inclined to claim that such extermination would be a paradigm example of an injustice. If a theory of justice easily allows the opposite to be asserted, then that theory must be objectionable. Barry, of course, assimilates this reaction to that of the defenders of Thomism cited earlier – the view of injustice here is held to be dependent on a substantive, ecocentric, theory of the good and thus will not commend itself to those who do not share that view.

In reply to this claim, it can be argued that the sense of injustice which the extermination of other species arouses does not necessarily rest on a substantive theory of the good, although that is a possible source. Rather, it rests upon another idea which hardly figures at all in Barry's account of his theory, namely the idea of the community of justice. The basic claim to be considered is that we should expand the idea of the community of justice to include at least some elements of non-human nature (precisely which these are is going to be a matter for further discussion). Once we accept that the interests and needs of non-human nature should be represented in the formulation of the basic structure of impartial justice – for example, in constitutional provisions, such as the right to environmental support for all members of the community of justice – then their extermination, including that produced indirectly by habitat destruction, will prima facie have to be regarded as unjust.

It thus needs to be shown how the idea of the community of justice can be detached from any substantive conception of the good. If this can be done then the basic claim which Barry puts forward for justice as impartiality, namely that it is independent of a substantive theory of the good, can be retained, although it will not have the force which he believed it to have. In particular, substantive disputes about the membership of the community of justice will be beyond the scope of justice as impartiality to resolve.

Why, though, accept the idea of justice as impartiality at all, and not simply go along with the idea that environmentally concerned people should be happy to espouse a conception of justice which is subordinate to an appropriate theory of the good, albeit at the cost of forfeiting any hope of gaining universal acceptance for it? The answer to this question is that Barry's theory does appear in many ways to be the best available theory of justice. It produces an excellent rationale for the kinds of outcome with respect to the purely human sphere which are attractive, such as toleration and mutual respect, and it promises to justify redistribution within societies and redistribution across them, as well as to underpin democratic procedures which environmentally concerned people will wish to support. However, to vindicate this claim would require a fuller defence of the theory of justice as impartiality than there is scope for at this point. This claim will, therefore, simply be taken as established for the purpose of this discussion.

The community of justice and justice as impartiality

We need, then, to show that the issue of the community of justice has to be addressed as a key part of a theory of justice as impartiality. As Andrew Dobson has written, the issue of how the community of justice should be constituted has scarcely figured in most mainstream discussions of distributive justice (Dobson 1998: 64). In elucidation of the concept, he cites Wissenburg, who has argued that the community of justice comprizes all those beings who are involved in distributive justice as dispensers, recipients, or both (Dobson 1998: 190–3, citing Wissenburg 1998). Clearly, human beings typically figure in both roles. However, some human beings, such as those with certain kinds of handicaps, will figure only as recipients.

The question of who should be admitted to the community of justice under the theory of justice as impartiality is not addressed by Barry to any significant degree, although he does say some things apparently germane to the issue, as we will shortly note. This neglect of the issue arises, I suggest, because of the way in which he sets up the problem to which justice as impartiality is supposed to be the solution. The problem is said to be that of determining how people with competing conceptions of the good may be brought freely to live together in peace on terms which all can accept as reasonable. This appears immediately to rule out, for example, non-human nature from the problem, and thus from the solution. For, obviously, no part of non-human nature, not even the intelligent primates, can be regarded as holding a 'substantive theory of the good' or be subject to the agreement motive. On this understanding of the problem they can appear only as items for consideration within some substantive views of the good, such as eco-centrism, held by only some people. Let us use the term 'primary group' to refer to the people picked out as comprizing the community of justice within Barry's formulation of the problem.

The importance of the 'primary group' to Barry's theory can be seen if we turn to consider his reflections specifically on the issue of our obligations to future human beings. He clearly believes that future generations cannot properly be dealt with on the contractual basis underpinning his version of justice as impartiality. Instead, he thinks that the latter should be treated only as having implications for, rather than as clearly establishing, our responsibilities towards future human beings (Barry 1998: 242). In the course of a recent discussion of the issue, Barry elaborates his views on the ways in which justice between contemporaries differs from justice between generations (Barry 1999). He suggests that the axiom of the fundamental equality of human beings which underpins justice as impartiality has corollaries which in most cases apply fully only in the case of contemporaries. However, he also suggests that the corollaries have important implications for how we should act in justice towards future generations, and there are in addition, he argues, connections between intercontemporary and intergenerational justice.

However, the discussion makes no overt reference to the basic contractual idea of justice as impartiality. It looks as though the main justificatory work is here being done by the axiom of equality. If so, then two problems arise. Firstly, alternative axioms will give different results. Secondly, axioms are not themselves supported by arguments. Barry expressly tells us that he knows of no way to support by argument the principle of equality of human beings (Barry 1999: 97). This implies that important conclusions about justice, those concerning future generations, rest directly on a basis for which no argumentative support can be given.

In addition, Barry concludes that the fundamental requirement of justice with respect to future generations is to maintain 'equality of opportunity' between them and us. However, since this conclusion emerges from the teasing out of the implications of the axiom, not from a Scanlonian contract situation, it does not lead to any further conclusions about either constitutional arrangements to produce it or forms of procedural justice which enhance the possibility of attaining it.

For both of these reasons, finding a way of incorporating issues of justice for future generations within the contractual apparatus of justice as impartiality appears advisable. This in turn requires that membership of the community of justice be broadened beyond the 'primary group'. Defenders of the idea of ecological justice who wish to marry their concerns to Barry's theory will also wish to rework that theory so as to admit even more members to the community of justice. How should they set about doing so?

Firstly, let us note that Barry himself realizes that justice as impartiality has to have something to say about those human beings who cannot be regarded as members of the primary group, namely those people mentioned in the 'marginal cases' already alluded to – people who have certain kinds of congenital handicap, infants, the demented, and so on. These are going to be members of the society created by the Scanlonian contractors, but are not members of the primary group. He castigates the rivals to justice as impartiality – 'justice as mutual advantage' (Hobbes, Hume, Gauthier) and 'justice as reciprocity' (Gibbard) – because, among other things, they cannot treat people in such categories as subject to justice. From the point of view of such theories, such people do not have the right kind of 'clout', or the ability to offer quid pro quos, to figure in the business of justice. This is because these rival theories view justice as concerning the ways in which people with power to affect each other's well-being adversely can be given self-interested motives to obey rules from which all benefit, even if some benefit more than others. Barry's comments on Gauthier's version of justice as mutual advantage provide a clear example of this line of criticism:

> Suppose we allow Gauthier the constraints that he artificially imposes, the range of justice is still by ordinary standards very impoverished. In particular, the 'congenitally handicapped and defective' fall outside its protection, because 'the disposition to comply with moral constraints . . . may be rationally defended only within the scope of expected benefit', and nobody can expect any benefits in return for maintaining them . . . I do not believe . . . that when we talk about enabling seriously handicapped people to lead productive lives we mean anything except enabling them to lead lives which are worthwhile to themselves. We are not suggesting that caring for them can

be shown to make a profit. That Gauthier thinks we must be claiming this simply shows that he seriously believes other people to occupy his own morally pathological universe.

(Barry 1995: 42)

However, it is far from clear how such people are supposed to receive protection within the theory of justice as impartiality, especially as they do not qualify for membership of the 'primary group'. Let us call such people the 'inarticulate'.

One possible immediate block to the claim that Barry's theory must admit into the community of justice members drawn from beyond the primary group is to argue that justice as impartiality is essentially about the primary group alone. This claim may be supported on the basis that the impartiality the theory propounds is precisely impartiality between rival views of the good. Thus it must concern only beings who have such views.

However, this is an incorrect interpretation of Barry's theory. For clearly it is intended to cover human beings who are capable of holding, but do not actually hold, a substantive view of the good. After all, not everyone does hold, even implicitly, anything that can be properly called a view of the good. Many people are in fact searching for such a thing. Barry notes that 'most' of us have a conception of the good 'of some sort' and that 'it will almost certainly not be fully articulated' (Barry 1995: 23). Even this cautious claim probably presents too sanguine a picture of the human condition. What Scanlonian contractors will nevertheless have, even if they do not actually hold a substantive view of the good, is interests of various kinds, including the interest in being able to pursue any substantive view of the good which they may eventually hold. Hence, it is such interests which Scanlonian contractors are fundamentally concerned to protect.

Clearly, however, those whom I have dubbed the 'inarticulate' can also intelligibly be deemed to have legitimate interests, even if they are incapable of ever entertaining a view of the good. Hence there is no reason deriving from the emphasis on impartiality for regarding Barry's theory as necessarily committed only to protecting the interests of members of the primary group.

To turn then to the issue of how such interests may be represented within the Scanlonian contract situation, one apparently plausible line to take is to suggest that, if the members of the primary group are given a veto over unreasonable constitutional proposals with respect to their own interests, then they will thereby so act that the interests of the inarticulate will also be protected.

However, it is hard to see how this outcome can occur on Barry's account of the constitutional provisions which emerge in the contract situation. These, as we have seen, contain protections for the exercize of religious belief and sexual preference and set out the requirements for procedural justice among those competent to take decisions on public policy issues. This provides no guarantee that the interests of the inarticulate will receive adequate protection. They might receive it if actual proxies are willing to come forward on their behalf in the democratic forums mandated by the theory, but that is a contingent matter. The situation which came to light in 1999 in that apparently justly constituted country,

Sweden, of the enforced sterilization, until only a few years ago, of those deemed socially undesirable, shows how easy it is for the interests of those without a voice to be suppressed in the service of certain views of the common good (*Keesing's Record of World Events* 1999: 42863).

Another suggestion to enable Barry's theory to handle the claims of the inarticulate is to argue that, from within the Scanlonian framework, members of the primary group would be likely to consider the possibility that they themselves may fall into the inarticulate group (for example, they may suffer dementia) and seek to protect their interests in such an eventuality. But that can go only a part of the way needed. It will not offer any protection for foetuses, for example.

A more promising line is to suppose that members of the primary group envisage themselves as likely to become directly involved in the well-being of at least some such inarticulates, for example in virtue of personal relationships with them. On this supposition it may be argued that they will have good reason to protect the interests of the inarticulate by constitutional bans on, for example, such practices as enforced sterilization. This will certainly cover many more relevant cases of inarticulates.

But this is all somewhat strained in terms of the basic motivation which Barry stipulates is possessed by the Scanlonian contractors, namely to find principles which are reasonable because fair. The strain arises because we are trying to find ways of defending the interests of people outside the primary group on the basis of the defence by members of the primary group of their own interests. But once we have imported the idea of fairness into the situation it is clearly more natural to consider the interests of the inarticulate directly, and to seek to deal with them fairly, whatever their relationship with members of the primary group. Barry correctly makes much of the difference that is made to contract situations when the motive of fairness is attributed to the contracting parties (Barry 1995: 50–1). But arguably his own conception of fairness is deficient, since it covers only fairness between members of the primary group. This is revealed in his comment (ibid.: 51) that the game structure of justice as impartiality is that of the assurance game, with each contractor doing what justice requires provided enough of the others do likewise. But if we are to suppose the contractors to be motivated by fairness in this respect, we cannot suppose them to be blind to the demands of fairness with respect to those individuals who are incapable of playing their part in an assurance game, but who have claims of justice against those who can.

The obvious way to achieve fairness for the inarticulate within the Scanlonian contract situation, in which we are imagining that moral debate is to take place between would-be contractors, is to require that the inarticulate receive the protection of their interests via proxies. Of course, a moral theory which did not employ the contract device would not need the device of proxies. Such a theory might simply argue directly for the moral importance of various kinds of inarticulates. However, then the proponents of the theory would themselves be acting as proxies. In the Scanlonian contract situation, the proxies will be members of the primary group who are willing and able to act as the guardians of the interests of the inarticulate. It should immediately be noted that such proxies are not to be understood as simply expressing their own preferences for the way in which inarticulate human beings are to be treated. Nor do we need to suppose that such proxies value or 'identify' with their 'clients', any more than lawyers do with theirs. They need only have a knowledge of their clients' interests, and a concern to defend them effectively in appropriate forums, in the interests of fairness.

They will, of course, be required to give justifications for the claims they will make about the interests of the inarticulates for which they are acting as proxies, as well as having the right to seek such justifications from others. In this way the idea of 'reciprocal justification' which lies at the heart of the idea of a community of justice remains operative. It is important to make this point in response to a view such as that of John Horton that 'the very conception of a moral community of justice seems to be radically transformed when it is accepted that many members are only recipients and not agents.'[1]

Thus, a defence of the introduction of proxies can be given within the terms set out by Barry for justice as impartiality, provided that we ponder the implications of the key idea that people wish to be given grounds that are 'not reasonable to reject and so fair' for accepting the basic rules of the society under which they are to live. The case of the 'inarticulate' enables us, plausibly, to suggest that what people accept in a Scanlonian contract situation as 'not reasonable to reject and so fair' is to be determined not solely by the question of how they are themselves to be treated, but also by the issue of how people not able to defend their interests are to be treated.

This then raises the possibility that some people will be concerned with the interests of another 'inarticulate' group – the non-human. Their concern cannot simply be excluded as inappropriate to issues of justice as characterized by justice as impartiality, since that is question-begging. If we wish to argue against it then we will need some further reason for refusing to admit the non-human to the community of justice. The fact that they cannot defend their own interests, be members of the primary group, or reciprocate moral concern will no more justify excluding them from the community of justice than it will justify excluding 'inarticulate' humans who are similarly situated. By now, of course, we are also aware of the various considerations advanced in the earlier chapters for claiming that all other non-human organisms are morally considerable, albeit to different degrees, and for saying that human beings are in circumstances of justice with respect to them.

All of this implies that in deciding the membership of the community of justice we are immediately plunged into a matter of controversy. As we saw in connection with Barry's criticisms of justice as mutual advantage, controversy on the issue of who to admit to the community of justice is present even when it is human beings alone who are being considered. In addition, many theorists, including Barry himself, will wish to refuse non-humans' admission to the community of justice even as recipients, which is the only conceivable role for them. In the case of Barry, his main claim in this regard, although it is not defended, is that relations of justice can obtain only between beings which are each other's moral equals, and this relationship of equality does not obtain between human beings and non-humans on this planet (although, to repeat, he does think that non-humans count for something, morally speaking) (Barry 1999: 95).

Of course the Scanlonian model which Barry employs rests squarely on the assumption of the moral equality of the Scanlonian contractors, and thus Barry's support for the assumption is not arbitrary.[2] However, the full adequacy of the model to our considered views about justice is precisely what is under discussion. The Scanlonian model seems to be attractive to Barry in part because it captures what is, for him, central to the idea of justice, which is that it obtains only between moral equals. But the latter claim needs independent support, which Barry has not (yet) provided. To say that it is a presupposition of the Scanlonian model is clearly not to offer such independent support.

Others who have discussed this matter, such as Wissenburg, see no problem in admitting non-humans as recipients of justice even though they are clearly understood not to have moral equality with human beings (Wissenburg 1993: 11). Wissenburg's view has recently been endorsed by Andrew Dobson in the course of a discussion which both highlights the importance of the issue of who gains admittance to the community of justice and marshals the arguments in favour of admitting non-human nature (Dobson 1998: 166–94). However, there is a key difference between his position and the one being defended in this book. This is that he accepts the location of the issue in the area indicated by Barry and Bell, namely within a substantive theory of the good (ibid.: 201–3). By contrast, we are here seeking to locate it within the foundations of a theory of justice as impartiality.

Before briefly outlining the arguments for admitting at least some of the non-human to the community of justice, let us take stock of the hypothetical problems created for Barry's theory by the existence of such controversy. That there are such disputes over who to admit to the community of justice may prompt various responses, some of which involve the rejection of Barry's theory, others of which do not. Thus, it might be argued that such differences over who should be admitted to the community of justice turn precisely on substantive theories of the good. If so, then obviously Barry's theory will have failed in its task, which was precisely to find a theory of justice which could prescind from disputes about substantive theories of the good.

Alternatively, it may be claimed that, whereas Barry's version of the community of justice does not rest on such a theory, the views of, say, defenders of the idea of ecological justice do rest on such a theory, namely ecocentrism. But this defence is incoherent, for his views and theirs are on the same logical level. If a theory of the good is presupposed by one such theory it must be presupposed by the other.

The final option is to argue that differences over the appropriate constitution of the community of justice do not turn on different views about substantive theories of the good at all. This rescues the basis of justice as impartiality, but shows that not all substantive disputes over moral issues inherent in establishing a theory of justice as impartiality are left behind when we prescind from substantive theories of the good.

This is the option which will now be explored further. It involves the claim that the arguments for including at least some elements of non-human nature within the community of justice do not depend upon the acceptance of any particular substantive theory of the good, but rather turn upon the characteristics of candidates for admission. On this view the key issue, explored earlier in this book in connection with DeGrazia's ideas, is whether such candidates can be said to have interests, or conditions in which they flourish or suffer harm. As we have already discovered, the issue of precisely how to use these conceptions across the whole range of non-human nature is a matter of complex debate. As we saw, many would argue that only some individual animals, possessing advanced forms of consciousness, can even prima facie be spoken of in these terms. By contrast, we have argued for the extension of these concepts to non-sentient life-forms. Others have argued for the encompassing of holistic entities, such as species, ecosystems and the earth's biosphere as a whole, within the community of justice. We will see in the next chapter that there is good reason to go a certain way in this direction, by allocating some claims in justice to collectives, rather than to individuals. We will not, however, find reason to go beyond this to embrace fully holistic entities of the kind just mentioned.

The only sure way of disallowing any of these moves is to restrict moral status to members of our own species. But, as is abundantly clear by now, it is extra-ordinarily difficult – probably impossible – to do so in a fully reasoned way. Most people, including, as we have seen, Barry himself, are prepared to accept that some non-humans are the proper objects of moral concern for their own sakes. Once this is accepted, however, the vocabulary of justice appears to be intelligibly applicable to them, as Dobson has noted in his recent thorough discussion of this issue (Dobson 1998: 68). For defence of their limited moral standing then appears to accord them the kinds of interests – for example, in avoiding suffering, having access to what they need for health and well-being – which give the concept of distributive justice its purchase.

It might be argued at this point that the idea that a being has interests which make it a candidate for membership of the community of justice necessarily involves the notion of what is for the good of such a being. Hence, the attempt to divorce the issue of who to admit to the community of justice from the issue of what substantive theory of the good to hold with respect to the candidates necessarily fails.[3] However, this argument should be rejected. Certainly, the idea that a being has the kinds of interests which make justice applicable to it does imply that it is the kind of being of which we can say that it has a good. But what this implies is that a set of indeterminate categories, such as mental and physical well-being, are meaningfully applicable to it. It does not imply any substantive view about what is conducive, or essential, to that being's mental or physical well-being. Hence the question of which substantive view of the good correctly applies to that creature is left open by the decision that it qualifies for membership of the community of justice.

Thus, even if it is accepted that at least some aspects of non-human nature are to be admitted to the community of justice under justice as impartiality, a whole set of issues concerning what is for the good of members of that community remains to be addressed. For the reasons given by Barry in his discussion of scepticism with respect to substantive theories of the good, it is unlikely that we will ever achieve unanimity on what is for the good of any member, or kind of member, of the community (Barry 1995: 168–72). But even if we had agreement on this issue there would remain the crucial further issues of how their competing interests, especially those of non-human nature and human beings, are to be reconciled.

However, although the decision concerning who to admit to the community of justice leaves open the question of what substantive theories of the good to adopt with respect to them, some moral positions are ruled out by the admission of non-human organisms. Obviously, those which accord no moral weight at all to non-human organisms are excluded. We need to be aware, however, that, from the point of view of Barry's theory, the moral position thus ruled out cannot itself be taken to be, or to presuppose, a theory of the good. To see this, we have to note that Barry is adamant that his own theory rules out the complete withholding of moral weight from members of ethnic or racial groups, even though some racist views would require precisely that. As we have already seen, this is because he claims that the theory of justice as impartiality presupposes that all human beings possess moral equality (Barry 1995: 8). This means that, a fortiori, it also presupposes that all human beings count, morally speaking. Since he supposes his theory to be independent of any substantive theory of the good, that implies that he believes that this 'all human beings count morally' thesis is also independent of any substantive theory of the good. But then he is logically required to grant that this applies to any other thesis about the moral considerability or otherwise of candidate members of the community of justice, such as those propounded by defenders of ecological justice.

Of course, Barry may be incorrect in maintaining the independence of the 'all human beings count morally' thesis from a substantive theory of the good. If he is, then of course the whole theory of justice as impartiality collapses. However, there is no reason to suppose that he is not correct and that the issue of which beings to admit to the community of justice is independent of any substantive theory of the good for them.

However, the issue of what moral weight to give to the interests of the different members of the community of justice does involve a substantive view of the good.[4] It has long been clear that, even when we admit non-human organisms to the community of justice, there is more than one possible view of moral weight between which to choose. One possibility is that canvassed by proponents of 'deep ecology', who argue for complete equality of moral weight between all organisms (see Doyle and McEachern 1998: 38–40). Another possibility, for which I have argued earlier in this book and elsewhere (Baxter 1999: ch. 5), is for a weighting system which seeks a principled way to give the interests of some organisms more weight than those of others, with certain kinds of human interests having the greatest weight of all. We saw in chapter 5 in connection with Wetlesen's theory how one might produce a reasoned argument for this view, and thus what the reasons are for rejecting the egalitarian approach. But it is clear from that discussion that substantive value-judgements of the kind involved in holding a view of the good are involved in arriving at that position.

However, if we admit this, have we not just given a proponent of justice as impartiality a reason to resist the admission of non-humans to the community of justice? The answer to this is negative, because, as has become clear from our earlier discussion, the same weighting issues arise when we restrict the discussion to the purely human case. Human inarticulates, such as foetuses, may be deemed in some liberal societies to have interests which are less weighty than those of fully developed moral persons, such that a clash between them should always be resolved in favour of the latter. In other liberal societies a different view may be taken. These are substantive value conflicts from which justice as impartiality cannot prescind. There is also, of course, the crucial issue of the point at which even human beings count morally in the course of their development from fertilized ovum, to which different liberal societies have given different answers, as has the same liberal society at different times.

These points emphasize that ecological justice does not introduce new kinds of issue into the basic levels of moral debate in human societies. They also show that the crucial 'inclusion' and 'weighting' issues involve the ineradicable presence of fundamental disputes over moral matters even from within the framework of justice as impartiality. 'Inclusion' issues raise such matters without involving a substantive theory of the good. 'Weighting' issues do involve such substantive theories. The latter may be incorporated into the liberal framework of justice as impartiality as matters to be dealt with by the procedural means allowed for within the theory, which appears to be acceptable and coherent. After all, actual liberal societies have to decide the precise forms which their constitutional provisions are to take, and employ the procedural means of democratic debate to do so. This is a matter of finding a determinate form for determinable general principles. As we saw when discussing Bell's arguments, such procedures are also what can be used to decide the 'inclusion' issues, and are so used in actual liberal societies to decide when a human being has entered or left the community of justice. But, given the substantive nature of the disputes involved, this means that these two kinds of issue attest to an ineliminable element of substantive moral dispute, even within the framework established by justice as impartiality and even when the discussion is limited to human beings.

What can now be said is that the proxies who seek to defend the interests of inarticulates, human or non-human, within the contract situation do not need to suppose that those interests always rank equally with those of other members of the primary group, or of other categories of the inarticulate. They need to be concerned only that the interests of the category for which they are acting as proxies are taken into account and accorded the appropriate weight. This begins to indicate how the idea of fairness inherent in ecological justice might be interpreted when non-humans are brought into the picture, a matter to which we will return in Chapters 9 and 10.

This latter suggestion may prompt a question directly relevant to the position developed here. It may be asked in response to the 'graduated weighting' proposal what the point is of seeking the admission of non-human nature to the community of justice if one simultaneously deprives its interests of equality of moral weight?

Doesn't this imply that the claims of human beings will always trump those of non-human organisms? Has the position of non-human organisms been in any way strengthened by this move?

In reply, it must first be noted that no emendation of a moral theory can guarantee even in principle the preservation of non-human nature. To meet a criticism made by John Barry of many non-anthropocentric moral positions, it must be emphasized that it is not being argued here that the vindication of the concept of ecological justice is required in order to guarantee a beneficial outcome for non-human nature (Barry, J. 1999: 58). Even the morally egalitarian position of 'deep ecology' does not guarantee that human interests will not trump non-human ones. What the amended version of justice as impartiality, which admits non-human nature to the community of justice, can plausibly aim for is to push moral thinking in a certain direction – one which requires the interests of non-human nature to be considered in human policy-making, which underpins constitutional provision for this, and which allows human interests to trump those of non-human nature only under certain fairly stringent conditions. Hence, the case for saying that extermination of other species is compatible with justice ought to become much harder to sustain than is the case with Barry's theory.

The upshot of all this is that the full elaboration of a theory of justice as impartiality may not produce a theory which can do what Barry wants. That is, even accepting the idea that the problem which justice as impartiality is designed to solve is precisely as Barry describes it, it will still remain a matter of controversy how to interpret the key ideas of what is 'reasonable because fair' and the 'agreement motive', precisely because how these are to be interpreted depends on the prior issue of who to admit to the community of justice and the issue of whether or not to attach differential weighting to the interests of some members over those of others.

Hence, detaching justice as impartiality from any substantive theory of the good and arguing that it presupposes the agreement motive do not guarantee that the nature-excluding version of justice as impartiality put forward by Barry is the 'only hope' for people to live together peaceably on the basis of what all can accept as a 'not reasonably rejectable and thus fair' solution. The most Barry can say is that, among people who wish to restrict the community of justice to human beings (or persons), his version of justice as impartiality is the only such hope, although even there the differences over weighting issues for at least some of what have been called 'inarticulates' involve substantive issues of the good which the theory appears to be unable to get around.

It might nevertheless be claimed that the versions of justice as impartiality of the sort offered by Barry, Rawls or Scanlon are the only versions of justice as impartiality which have full coherence. Arguably, the incorporation of ecological justice into a theory of justice as impartiality involves such a radical transformation of the latter that its underlying motivation and structure disappears.[5]

After all, in the course of the discussion it has been proposed to extend the constituency of justice beyond the paradigm contractors of the primary group; to broaden the scope of fairness beyond the interests of the primary group members;

to insist on a distinction between agents and recipients of justice; to admit as members of the community of justice entities incapable, even in principle, of moral thought; to admit inequalities of moral weighting between different elements in the constituency of justice; and to admit proxies into the contractual situation. Does the underlying idea of a contractual situation survive all these modifications? If it does not, does anything survive of the idea of justice as impartiality?

In reply, one can suggest that the most basic element of the theory and its contractarian core does remain, namely the idea that 'justice as impartiality is fundamentally about justifying to each other the terms on which we can live together.'[6] The position defended in this chapter is intended to suggest that this justification can and should go beyond what affects the interests of the members of the primary group, even though these are the only beings who can ask for, and be owed, a justification. But the usefulness of the contract idea is not itself impugned by this more complex tale.

In addition, the heart of Barry's version of the theory remains intact – for impartiality at the most fundamental level between rival conceptions of the good (so it has been argued) remains possible. This is a significant result even if the admission of inarticulates, even solely human ones, to the community of justice raises ineliminable issues of weighting which involve substantive views about matters of fundamental value. What has also been argued, however, is that this mode of impartiality, important as it is, is of much less significance for arriving at a justifiable theory of justice than Barry's discussion would have us believe.

Since, then, there is at least one, rival, ecological version of justice as impartiality, there arises immediately a new version of the issue which Barry's theory was in large part intended to solve, namely how peaceful coexistence may be obtained between its supporters and those of Barry's version. It looks as if there are only two possibilities. One is for it to depend on the purely prudential basis that open conflict between them is worse than mutual toleration – a solution which Barry had hoped to avoid, because it involves an inherently unstable situation, with each side remaining ready to seek the opportunity to overthrow their opponents (Barry 1995: 39).

The other is to find a 'meta'-theory which does for rival versions of justice as impartiality what justice as impartiality was intended to do for rival substantive theories of the good. What this might be is at present hard to see – perhaps it is even impossible. For what this discussion may reveal is that the key aim of justice as impartiality is unobtainable – namely to secure a theory of justice which all can agree to, whatever their views on substantive moral matters.

9 Ecological justice and the non-sentient

The position which we have now reached is as follows. We began by seeking to vindicate a universal approach to moral theory as against the contextualism which has become popular among some proponents of environmental ethics. This was necessary in order to permit the possible development of a moral theory of the relations between human beings and the non-human beings with which we share the planet which, if valid, would be meaningfully addressable to all moral agents, whatever their cultural context. We then spent time attempting to defend the view that non-human organisms are all morally considerable. Specifically, we aimed to show that there is no good reason to exclude the class of non-sentient organisms from that category. We endeavoured to defend the reasonableness of a theory of differential moral weight as applied to various kinds of organism. We then argued that our moral responsibilities towards non-human organisms are not simply those of humaneness, but encompass requirements of distributive justice, in spite of what appear to be insurmountable objections to the idea that beings which are solely moral patients can be proper recipients of such justice. We then examined some eminent theories of distributive justice with respect to the purely human case to ascertain whether any of them could accommodate the idea of distributive justice towards the non-human. Two candidates proved to be incompatible with such a notion. However, latterly we sought to show that the liberal theory of justice which has emerged from the Rawlsian matrix is capable of encompassing justice towards the non-human and of justifying some constitutional provisions to secure the claims of the non-human to their fair share of environmental resources.

In the remaining chapters we turn to the difficult issue of trying to specify what distributive justice towards other organisms involves. This will require us to formulate and justify some specific claims with respect to such theoretical issues as:

1   Which entities are the bearers of the claims in justice to environmental resources?
2   To what resources may they properly lay claim in the name of justice?
3   How are moral agents who seek to meet the proper claims to resources of different claimants to adjudicate disputes between them?

We will also need to say something reasonably specific about the institutional arrangements that will be required within and between human societies for the protection and implementation of ecological justice. This will involve in part a discussion of 'ideal' theory – what institutional arrangements would ecological justice require if they could be established by fiat – and in part a discussion of how proponents of ecological justice may reasonably hope to achieve their aims in the complex world we actually live in.

Let us now turn to the first set of issues. We will consider the case of non-sentient organisms in this chapter, and sentient ones in the next. It is a matter of some controversy as to which kinds of organism should be regarded as possessing sentience, understood as the capacity to feel pleasure or pain. Mary Ann Warren distinguishes sentience from consciousness, on the basis that not all conscious experiences are pleasurable or painful (Warren 1997: 55). But, as she goes on to note, 'It seems likely that most naturally evolved organisms that are capable of conscious experiences are capable of experiencing (among other things) pain and pleasure' (ibid.). This looks to be plausible, and suggests that the sentience/consciousness distinction may be hard to make much use of in practice.

Sentience seems on most informed views to be clearly absent from organisms which lack any kind of nervous system or sense organs, such as plants and single-celled organisms. But there is room to doubt that even when an organism does possess some sort of nervous system it thereby necessarily possesses sentience (Warren 1997: 62). What specific organisms may be said to possess sentience is clearly a matter of ongoing empirical investigation and theoretical advance.

Let us now consider what may be said about the moral claims of the members of the non-sentient group of organisms, the 'merely living', however we specify them in practice.

Right-bearers and the 'merely living'

The first issue to consider is what are the appropriate bearers of claims to environmental resources in the case of the 'merely living'. There appear to be only three possibilities: individual specimens of a species; specific populations of a species; and the species as a whole. Let us examine these in turn.

Arguably specimens of the merely living are too lacking in individuality for it to make much sense to attribute the rights to individuals of the species. This is because no individual specimens of these species possess particular ambitions or purposes, different from those of any other member of the species, which would make possible a meaningful moral differentiation between them. The killing of such specimens thus thwarts no plans or projects unique to the individual. The projects of individuals are interchangeable, and so it makes straightforward sense to say that the individuals themselves are interchangeable or fully replaceable. The giving of moral priority to the group over the individual is, in their case, fully justifiable. Their individual welfare interests remain, of course, but are impossible to disentangle fully from the welfare interests of the group.

The individuals themselves will of course be numerically distinct and may be functionally differentiated within certain subgroupings, as in the case of workers and drones. They will also possess sufficient difference at the level of their genes to allow scope for the workings of natural selection, workings which can, of course, result in time in the emergence of more individualized creatures. In addition to this lack of individuality such non-sentients will, of course, be devoid of consciousness, sentience and self-consciousness. This means that there is no individual suffering to be weighed in the moral balance. We can express the upshot of all these points in a concrete manner by saying that swatting a fly usually need cause no moral compunction, although there still needs to be some defensible reason drawn from the interests of the swatter for doing so, such as the need to avoid the infections which flies unwittingly can bring into our households.[1]

If, then, we can justify the attribution of claims of the non-sentient to environmental resources to the group rather than the individual, leaving the individual with only a residuum of moral considerability, should we grant these claims to the species? This is a view to which I have given support elsewhere (Baxter 1999: 80). However, it is clear that this alternative also encounters serious problems. There is, first of all, the standard ontological problem about whether or not any entity to which the term 'species' can be applied really exists. This is not the same as the problem of whether or not species have determinate boundaries. Clouds undeniably exist, but we cannot say what their boundaries are with any precision. The problem is rather that the concept of 'species' may simply be a useful abstraction, a way of referring to creatures en masse for certain purposes of biological theory. If this is the correct view to take then it may be enough to deprive species of any moral standing. But it may not. After all, at least in the legal systems of many countries, commercial corporations are granted the legal status of personhood, and granted rights, even though such corporations are abstractions.

However, from the point of view of ecological justice, attributing rights to environmental resources to species in the case of the 'merely living' has the potential for an unfortunate practical implication. For it might have the effect of allowing persons to engage in large amounts of destruction of individuals and populations to the point where the minimally viable numbers of the various species are reached somewhere on the planet. To some this will be an acceptable conclusion. But for those who are motivated to produce a theory of ecological justice something will have gone badly wrong if this is the result of such a theory. What has gone wrong is that this approach goes too far in the direction of subordinating the individuals of such species to the species as a whole. Although we have seen that such species contain members which are not strongly individuated, this does not mean that such individuals are totally devoid of moral standing. Hence the attribution of rights to species in the case of the merely living is subject to a moral objection to which the practical implication just noticed alerts us.

There are two other, practical, concerns which the exponents of ecological justice will have with respect to the minimalist approach. The first is that there may be misjudgements about what constitutes a minimal viable population of any species, and that this level will be overshot, leading to extinctions. The second is that the task of protecting the minimum viable population will be left to others, or be mistakenly supposed to be undertaken by others, so that extinctions result from universal buck-passing.

We have seen, then, that with respect to species of the merely living there are serious conceptual, practical and moral problems with attributing natural resource rights either to individuals or to the species as a whole (and the problems with attributing such rights to species are also applicable to species which are more than merely living). For these reasons, therefore, with respect to the merely living, we should conclude that it is populations which are preferable as bearers of the right-like claims embodied in ecological justice. They unquestionably exist, and are usually distributed through suitable ecosystems so that granting them the right to the resources needed for continued existence and flourishing optimizes the chance that they will not be pushed to the brink of extinction. Since the reasons that have brought us to this conclusion apply also to those organisms which possess at least sentience, we may conclude that in their case, too, the attribution of morally defensible claims to populations is justifiable as a minimal position. That is, the least we can say is that the claims of ecological justice can properly be attributed to populations of such organisms. However, as we move beyond sentience to consider the individualizing characteristics such as self-consciousness and various forms of intelligence, it may well be necessary to move beyond this minimum claim. This is a matter to be investigated in the next chapter.

Taking this line does not involve us in the controversial speculation that populations are superorganisms, that is that they are themselves a form of living entity. Rather it involves pondering what is for the good of non-individualized (or minimally individualized) species. The answer to that can only refer to the existence of populations of the species above a certain threshold level, for it is within these that the good of the kind of organism to which we are referring resides. The welfare of such organisms is largely a matter of being members of populations sufficiently large to allow interbreeding and reproduction in sexual species. Membership of a population allows such behaviour as schooling to take place, which reduces the opportunity for predation on any given individual member (McFarland 1981: 490–4).

This leads to a further point about the good of such species which justifies us in attributing right-like claims to populations rather than to individuals. For many such species have members whose natural mode of life is to compete with each other for such resources – the phenomenon of territoriality (McFarland 1981: 551–6). Hence, in their case any claim against moral agents should be held by the population, for it will be a claim for sufficient resources within which particular members of the population can fight out their territorial battles in accordance with their natures. In this we have an echo of the Rawlsian principle that in a just society the aim of the just constitution is to enable members to compete for prized goods, such as positions and offices, under conditions of fair equality of opportunity.

This suggests one way in which individual members of populations of territorial species, both sentient and non-sentient, may be deemed to have individualistic right-like claims against moral agents, namely the claim not to be harmed in such a way that they lose the chance to compete for mates and territory. In intelligent species which are the subject-of-a-life this may lead to harms other than simply the chance to breed and eat; it may lead to psychological states which are bad, such as depression and frustration.

However, if we do take the view that, with respect to the merely living, it is populations which should be deemed to be the bearers of right-like claims for the purposes of ecological justice, then we need to allow that it applies only to viable populations. A population which is heading towards extinction in one locality through non-human processes arguably loses its right to environmental resources. We have to allow that both local extinction and total extinction of species is a part of the ongoing process of life. Just as, in the human case, we should not kill (unnecessarily) but 'need not strive officiously to keep alive', so we should not do so when a species is facing extinction as the result of irreversible changes in environmental factors which have not been produced unnecessarily by the activities of moral agents. Changes which are so produced, are, of course, subject to moral critique, on the basis of violation of the requirements of ecological justice, for example. However, we need to look out for self-serving human analyses of situations in which species are claimed to be near extinction as the result of unavoidable changes in the environment.

We also need to note that the concept of a 'viable population' is fuzzy. It is fuzzy in two dimensions. Firstly, it may be a difficult matter to specify what is meant by the term 'population' in a given instance. The concept is most clearly usable in a determinate sense when it is possible to establish precisely the boundaries of a number of organisms of the same species. However, in the case of some organisms their presence may be continuous throughout an extensive range – perhaps as large as a continent. If so, it will, of course, be hard to substantiate the claim that localized human activity has any clear effect upon the population in question. In such cases the issue may better be conceived of as that of whether human activity is reducing the species as a whole to some limit of viability. Of course, it is likely that, before that point is reached, human activity may succeed in creating gaps in the total range of the species, in which case it will probably become possible to speak once again in terms of the impacts upon particular populations.

The second dimension which may be fuzzy is that of 'viability', where it is likely that any specification of limits of viability will inevitably be presented in terms of a wider or narrower range. It will be important to emphasize, however, that the concept of viability which is employed in the context of ecological justice is one according to which the population is viable without continuous human intervention (through special breeding and release programmes, for example). If we were to allow human-assisted viability of this kind then we would be depriving the concept of population viability of any cutting edge, since any threatened population might thereby be deemed maintainable on a basis that might well be purely hypothetical and require levels of human involvement that might not be sustainable in the long run. It is important to note that this point does not rely upon an opposition between the human and the natural. That is, the objection is not that human-sustained viability is objectionable because 'unnatural', rather it is that it cannot be presumed to be automatic and so cannot be relied upon.

However, the attribution of rights to fuzzy groupings is not an insurmountable difficulty. After all in the purely human case we can intelligibly attribute rights to groups the boundaries of which may, at any given time, be ascertainable only within a certain range of indeterminacy. For example, we routinely attribute the right to elect the government to something called the electorate, even though it may be unclear who exactly qualifies to be a member of that body and there will be cases under dispute at any given election. The concept of the electorate will nevertheless be determinate enough for the purpose at hand, and be an intelligible bearer of rights.

A fundamental aim of the theory of ecological justice, therefore, is to defend the claim that viable populations of organisms which are 'merely living' have a prima facie right to environmental resources necessary for those populations to continue to exist in a way which permits the flourishing of their individual members (that is, the population must not be reduced to merely 'hanging on'). This conclusion, as we have noted, may be sustained by teasing out the idea of what harms such species and taking it beyond the purely individualistic level which prevents no more than thoughtless destruction of individual specimens. In case the foregoing reasoning might appear to be too divorced from the practicalities of the conservation of organisms, it is worth noting that in the preamble to the UN Convention on Biological Diversity of 1992 we find the following passage: 'The Contracting Parties . . . Noting further that the fundamental requirement for the conservation of biological diversity is the in situ conservation of ecosystems and natural habitats and *the maintenance and recovery of viable populations of species in their natural surroundings*' (Le Prestre 2002b: 345–6; my emphasis).

If this is a claim of justice, rather than of humaneness, this means that it can be overridden only by fairly weighty considerations. What duties, therefore, can we say are owed by human moral agents to a viable population of merely living organisms which have a claim in justice to environmental resources needed to survive and thrive? What is permitted and what is forbidden?

A system of moral trade-offs between the merely living and other organisms

One point which is immediately clear is that the claim in justice for a fair share of resources cannot be met if the moral agents whose duties are triggered by the right-like claims are unaware of the existence of the population. Hence, moral agents have the corollary duty to ascertain what populations exist within those environments within which they are proposing to operate, and to ascertain what are the needs for the continuation of viable populations. Many biological scientists, such as Edward Wilson, have argued that we urgently need an inventory of the species with which we share the planet (Wilson 1994: 297–305). These urgings have now begun to produce official action. For example, in recent years various international NGOs, together with environmental and other institutions have launched the Global Biodiversity Assessment, the Millennium Assessment of Global Ecosystems and the Global Biodiversity Information Facility (Le Prestre 2002a: 320). What the theory of ecological justice makes clear is that it is a matter of justice, not just of human interest or well-being, to discover not simply which species exist somewhere on the planet, but where their viable populations are – or, at least, how this may be ascertained in any given instance.

A further point we can make straight away is that we cannot simply exterminate a population of some merely living organism, directly or indirectly, on the basis that it is not from an endangered species. For it is the particular population which has the right-like claim of justice, not the species. It may be permissible to reduce their numbers to a level lower than they currently enjoy, provided they are left with a viable population. We may be required to consider ways in which their needs for resources can be catered for, a procedure which may require greater expenditure of time and energy on our part than would otherwise be the case. To put it in more general terms, it is clear that where a viable compromise can be reached between their needs and the needs of moral agents then this should be taken, even at a greater cost to the moral agents than would otherwise be the case. Where alternative resources may be made or left available, different from those which the species normally uses, but still usable, then this may be a required form of compromise.

Clearly what is here being proposed as the set of requirements upon moral agents intrinsic to the concept of ecological justice involves the imposition of very significant information and calculation costs on those agents. It will cost time and money to discover what species exist, what populations of these exist, how they are distributed, what their ecological and other biological needs are and how these all may be affected by various proposals for human beings to act in pursuit of legitimate interests and projects. It is precisely because of the size of the costs here that a concept as weighty as justice is needed to justify that expenditure.

However, there are various factors that can be mentioned which may have the effect of mitigating the apparent onerousness of the tasks being proposed by ecological justice. To begin with, a theory of ecological justice does not have to suppose that moral patients are helpless in the face of threats to their well-being from moral agents. Organisms can often adapt successfully to new conditions imposed upon them by moral agents, without this involving evolutionary changes in their structure or behaviour. Sometimes the changes may even be advantageous, enabling members of a species to do better than before. Of course, many species have co-evolved with human beings, and live in symbiotic or commensal relationships with us. For reasons such as these it may not in practice be necessary every time for human beings to do anything very much to protect the interests of some populations affected by human activities.

With respect to the other side of the coin, namely that as members of a biological species, human moral agents are subject to attacks upon their interests by species we refer to as pathogens, pests and vermin, ecological justice will seek to ensure that a proper attempt is made to guarantee that these organisms are not being incorrectly given an adverse valuation and that no modus vivendi can be struck between human moral agents and them.

However, the most fundamental task of a theory of ecological justice is to produce a set of principles which enables it to be determined in a non-arbitrary and morally defensible way how to trade off clashes of interests between different organisms in competition for the same environmental resources. Let us now consider how this might be done.

We have already noted, in considering Warren's theory, the idea that the working out in a given instance of how moral agents are to reconcile the claims of different affected organisms on the basis of their different levels of moral status is not a matter to be resolved by the application of a meta-principle or principles. This view is probably shared by Wetlesen, who offers some considerations to be adduced when we are considering alternative courses of action which will all have some negative impact on the interests of organisms, and we are trying to decide how to act. There is no clear decision procedure adduced in the course of the presentation of these considerations, nor is it clear that these are the only relevant considerations which can be referred to in such circumstances. For both him and Warren, therefore, the matter is irreducibly a matter of judgement. Presumably, therefore, reasonable moral agents may differ over how the situation is to be interpreted in the light of the moral status of the affected parties and the kinds of consideration presented. It will be helpful to have the list of considerations offered by Wetlesen.

1 What kinds of living organism are affected by the possible consequences of the alternatives of action we can choose among?
2 What are the probabilities of these consequences?
3 What kinds of interests do they affect – are they central (vital) or peripheral?
4 How many organisms are affected negatively or positively by each alternative?
5 What degree of inherent moral status value can be attributed to each species of organism affected?
6 Do any of these organisms belong to species or ecosystems that are ascribed a special value, either instrumentally or intrinsically, by other agents whom we should respect?
7 Are any of these species or ecosystems endangered?
8 Do any of these organisms belong to species or ecosystems in relation to which we have contracted special moral obligations through our earlier actions, for instance by domesticating them?

(Wetlesen 1999: 317–18)

As we noted earlier, Wetlesen's discussion is avowedly individualist and does not operate with any distinctive orientation towards matters of distributive justice between species. As a first step we might alter some of his considerations to convert them into ones relevant to such matters.

1a What populations are affected by the possible consequences of the alternatives we can choose among?
2 As above.
3a Do they affect the share of environmental resources currently enjoyed by the populations?
3b Do any of the courses of action which affect resource share take any of the populations below viability?

4a How many populations are affected negatively or positively with respect to resource share by each alternative?

5a Replace 'species' with 'population'.

6   As above.

7   As above.

8   As above.

The effect of these alterations is to insert a level of analysis above that which emerges in the account of Wetlesen and Warren. That is, whereas their concerns are directed towards elucidating the duties of moral agents with respect to individual organisms, the ecological justice considerations are directed towards their duty to respect the claims to resources of populations of organisms. This implies the possibility that on occasion there may be a clash between what has to be done to take account of any right-like claims of individual organisms against moral agents and what has to be done to take account of the claims in justice of populations of organisms to the resources they need for viability. When this occurs, given the earlier points about 'harm' and 'well-being' for merely living organisms being bound up with the viability of the local populations which they inhabit, ecological justice would demand that the requirements of justice for the populations should be given priority.

Of course, the individual organisms in question are, as earlier noted, devoid of much individuality, so that they may more easily be sacrificed for the common good of their group than is the case with maximally individualized organisms such as human moral agents. With respect to the latter it is normal for theorists to argue that there are certain sacrifices that may not be enjoined upon individuals in the name of the common good and that this restriction is a requirement of justice itself.

Further, and as already noted, it will also be the case that, as we consider the claims of organisms with greater moral status than the merely alive – those which are conscious, sentient, self-conscious, subject-of-a-life – the requirements of ecological justice will increasingly encompass the claims of individual organisms to their fair share of resources in order to accommodate the increasing degrees of individuality which they manifest, and thus the resources needed to pursue particular pathways through the space of possibilities open to them. This culminates in the fully moral agent case in which we have the statement of individual rights in its full, familiar, sense.

We noted earlier that the rights attributed by ecological justice to the merely alive species were strongly prima facie, that is, they can be overridden in specific instances to accommodate the rights of species with more moral weight. This is the reason why the adjudication of moral claims in specific instances cannot be reduced to the application of some meta-principle, such as that specifying lexical ordering of right(-like) claims. However, the idea of ecological justice is that these right-like claims should not be negligible, but have some clout in the course of moral deliberations. Is there any way of ensuring that the right-like claims of populations of the merely alive can be given weight in moral deliberations without resorting to a lexical ordering?

One possibility, for which I have argued (as have others) elsewhere, is that we should give different kinds of weight to different kinds of interests (Baxter 1999: 81). Wetlesen's list of considerations noted above makes a distinction between central and peripheral interests, but does not explain how it is to be used in the course of moral deliberation. An intuitively plausible approach is one based on a hierarchy of necessary conditions. Thus, for any organism, surviving is a necessary condition for thriving as an individual, which in turn is normally a necessary condition for reproducing itself. These can all properly be called basic interests, and all of them are weightier than its interest in short-term gratifications. Any organism has, therefore, a range of interests which are properly called 'basic' interests, although they can be arranged in an ascending order of weightiness.

What is notable is that the simpler an organism is – if, for example, it is 'merely living' – the smaller its range of interests and the more these coincide with its basic interests. An organism geared solely to surviving and reproducing itself is one in which all its interests are basic ones. As organisms become increasingly complex and acquire new properties, such as sentience, consciousness and so forth, the concept of flourishing for them becomes more complex too, as are the correlative harms to which they may be subject. But there is also an increasing list of interests which go beyond the 'basic' interest category.

The suggestion which may then be made is that the basic interests of a population of the merely living can only be overridden by the basic interests of populations of morally more weighty species. Their basic interests cannot be overridden by the non-basic interests of morally more weighty species. As we consider the clash of basic interests between more and less weighty species we will discover that the range of basic interests of the weightier species will become more complex and extensive – especially as involved in the notion of 'thriving' or 'doing well'. It will thus appear that the basic interests of the 'merely living' can rather easily be overridden by the basic interests of morally weightier species, and with supreme ease by the basic interests of species which are moral agents.

However, the increasing complexity of the content of 'basic interest' when applied to the thriving of morally weightier species also permits many more options to be available for the realization of those interests. For example, as I have argued elsewhere, it is a basic interest of moral agents who are persons to develop their talents and abilities to some reasonable (but perhaps unspecifiable) extent. But it is in the nature of many such talents that they are determinable rather than determinate, and thus they may be realized by a variety of life-projects (Baxter 1999: 154). It is arguable, and liberals standardly do argue, that the decision of how to develop one's talents should be left up to the individuals whose talents they are.

However, if the development of one's talents leads one to attack the basic interests of a population of merely living organisms, by depriving them of the share of environmental resources they need to survive and flourish, and it is possible for one to develop one's talents by an alternative which does not do that, perhaps on the basis that society could help one to find an alternative, then ecological justice would require that to be done. Your basic interests would still be being met, although not as easily, perhaps, as might have been the case if you had been allowed to override the basic interests of the population of merely living organisms in question. This does not require human beings to move away from areas where they may be attacked by organisms which threaten their life or well-being, for such areas are coextensive with the whole planet. In such cases extermination of specific organisms may be the only effective resort.

One might say, then, that the basic interests of persons are more extensive and more varied than those of less weighty organisms, which seems to threaten to justify quite easily the overriding of the basic interests of less weighty organisms. On the other hand, their very sophistication allows for many more alternative modes for their satisfaction, and thus more scope for compromises to be found when clashes of interest arise – compromises, of course, which put the onus upon the moral agents to look for alternatives. As earlier noted, this search for alternatives is not devoid of costs, which will always set up a pressure to ignore the basic interests of less weighty life-forms when they clash with those of moral agents. But the point, once again, of making this a requirement of justice is to put the weight of that concept behind the claims of the merely alive.

These points, of course, make plain that the issue of distributive justice between species, ecological justice, is inextricably entwined with the issue of distributive justice between persons. The issue of the distribution of benefits and burdens between humans (including environmental benefits and burdens) is not discussable without referring to its impact upon the claims in justice of non-person organisms. Two alternative patterns of distribution between which human beings may be indifferent if it is just human interests that are at issue may be very different in terms of their defensibility once the claims in justice of non-person organisms are introduced into the discussion.

A final implication of the position here being argued for has to be noted. This is that, when the choice facing a moral agent is that of sacrificing the basic interests of one or other of two non-sentients, each equally non-individualized, then, *ceteris paribus*, it is a matter of moral indifference which is chosen. However, morally relevant 'tie-breakers' can be employed in certain circumstances, such as those suggested by Mary Ann Warren and Jon Wetlesen, namely being members of an endangered species, or of a keystone species, or both. Such considerations may give the edge to one individual, in a morally relevant way, from the point of view of ecological justice. That is because the fact of being an endangered and/or keystone species has a bearing on the welfare interests of the organisms concerned and of others with which they are interconnected. In this respect these considerations are germane to ecological justice in the way that other possible properties would not be, such as being aesthetically pleasing or culturally significant to certain groups of human beings. This is not to say that such properties could not ever be morally important. But it does mean that they are not directly concerned with ecological justice, and it is justifiable to claim that matters of ecological justice should, prima facie, trump other moral considerations.

Ecological justice and other moral issues of human–non-human relations

Thus far the view which we have espoused concerning the goods to be distributed by ecological justice has focused exclusively on environmental resources. Are there any other kinds of good to which a theory of ecological justice ought to entitle non-human members of the community of justice? In the earlier version of Rawls's theory of justice the people in the Original Position focused upon a set of goods known as 'primary goods', conceived of as indispensable general means to any particular plan of life which they might wish to pursue when their concrete identities became apparent after the removal of the veil of ignorance. The two principles of justice to which Rawls's theory leads concern one kind of primary good – what he referred to as the class of 'social primary goods': rights and liberties, powers and opportunities. The use of this idea in Rawls's theory has been the subject of frequent criticism which need not concern us here. It is, however, an idea which might enable us to pose the above question more precisely. Thus, one might ask whether, in the case of non-human organisms, there are any primary goods, beyond those environmental resources upon which their existence and flourishing depend, with which moral agents ought to furnish them as a part of ecological justice.

This question is made complicated by the fact that, as we have frequently noted, moral agents are themselves biological organisms, and it is clear that they will often need to make use of other organisms for life-sustaining purposes. So far we have been arguing that ecological justice requires them to take strenuous measures to ensure that in the course of this they do not so act as to deprive any population of a species, without good moral cause shown, of the environmental resources necessary for the continued existence and flourishing of the members of that population in accordance with their own nature.

But human moral agents typically have many other effects upon other species than the extermination of local populations and ultimately whole species. Should not ecological justice theory say something definitive about such human–non-human relationships as meat-eating, domestication and selective breeding of other species, and experimentation upon their members for medical, scientific and agricultural purposes? As we noted right at the start, it is also not unknown for human persons to invest non-human species with personal, cultural and religious significance. Does this not mean that ecological justice will need to say something principled about pets and those species viewed as sacred or unclean, or as embodying national spirit (the bald eagle) or totemic significance, insofar as these valuations may lead human beings to destroy or protect creatures in ways which contravene the requirements of ecological justice?

Perhaps surprisingly, the answer to these questions is, from the point of view of ecological justice, in the negative. That is to say, provided that viable local populations of non-human species are not destroyed without good moral cause shown, and thus provided that existing species are left overall in a flourishing condition, it is not a matter of ecological justice to resolve issues of meat-eating, domestication, animal experimentation and so forth.

In justification of this standpoint we can echo the claim of Rawls that the theory of justice is not the whole of morality. There are undoubtedly moral issues concerning human treatment of non-human organisms which are not encompassed by considerations of fair shares of environmental resources. Causing unnecessary pain to sentient animals is a case in point. Thus, good moral arguments are available to demonstrate the prima facie case against all the ways in which human beings cause other sentient creatures unnecessary suffering, in zoos, on farms, on the hunting field, in laboratories, in places of entertainment and so forth. Key to these arguments will be the detailed demonstration of the kinds of suffering to which different types of animal are prone, of the sort so meticulously and convincingly provided by David DeGrazia. Certainly, too, exponents of ecological justice will be committed to taking such arguments with extreme seriousness, for both welfare and justice arguments with respect to the non-human rest on the prior acceptance of its moral considerability.

However, it is possible to be a supporter of ecological justice without thereby being committed either way on the issue of, for example, meat-eating. One might take the view that, on the balance of the argument, it is morally wrong to eat other animals because of the suffering they experience in the course of being raised and killed, or hunted, for food. One might employ arguments based on the notion of respect for the lives of individual animals. Or one might argue that, provided that no suffering is caused to animals, eating them for food is not in itself morally objectionable.

The strength of these arguments is going to depend very much upon the specific characteristics of the animals in question, and the more that animals embody elements of moral personhood the stronger the case against treating them in certain kinds of ways. This is a matter to be considered more fully in the next chapter. There are also moral considerations other than those which relate to animal welfare which may be relevantly brought in. For example, the raising of large numbers of animals for consumption has implications precisely for the habitats of many other species, as well as increasingly apparent implications for more general environmental phenomena, such as the contribution to climate change made by the release of the greenhouse gas methane by increasing numbers of cattle world-wide. In such ways an ecological justice argument may be brought in by an indirect route, for the advent of increasing numbers of domestic animals may threaten the continued existence both of local populations of species and of species as a whole. However, the deployment of that argument does not involve a decision on the rights and wrongs of meat-eating per se.

Something similar may be said with respect to other burning issues concerning human–non-human relations. The genetic modification of organisms to serve human agricultural, medical and scientific interests does not directly concern issues of ecological justice, provided that the creation of such organisms does not threaten the continued existence and thriving of local populations of species or their overall existence. However, there are many other kinds of moral consideration which have a relevant bearing on these activities, such as animal welfare ones (is any unnecessary suffering involved?), perfectionist ones (is the dignity of the creature involved attacked?) and all the usual issues surrounding the possible harmful effects on human beings of such practices.

We may conclude, then, that ecological justice, as only a part, albeit a crucial one, of the general issue of the morality of relations between human beings and non-humans, does not in itself have any implications for these other questions. If, however, they can be shown to impact on the continued survival and flourishing of local populations of non-human organism, and thus potentially on the continued existence of other species, then ecological justice is involved in their resolution. To put the point in terms of primary goods, we can therefore say that the only primary good at issue in the topic of ecological justice consists of the environmental resources needed for populations of species to survive and flourish after their kind.

Let us next examine whether any further refinements to this general position emerge when we consider other species which are at least sentient and which may possess consciousness and self-consciousness and various forms of intelligence.

10 Ecological justice and the sentient

It is important to emphasize at the start of this chapter that, although we are now moving beyond the non-sentient realm of the 'merely living' to organisms which possess sentience, consciousness, self-consciousness and various forms of intelligence, our focus is still firmly upon the requirements of ecological justice and upon what environmental resources such organisms are entitled to claim against moral agents. We are not concerned with the issues of how moral agents may treat or mistreat such organisms in other respects, such as in their direct personal inter-action with them. It is important to emphasize this point because it is with respect to organisms possessing sentience and beyond that an enormous amount of ethical thought has already been expended without, however, addressing issues of distri-butive justice regarding environmental goods and bads of the kind with which ecological justice is concerned. We need to maintain, therefore, a firm grip upon this set of issues, without, of course, disparaging the efforts of those whose concern has been primarily with the welfare and suffering of individual non-human animals.

What are the appropriate subjects of claims to resources as we move to the class of sentient organisms, and to what does ecological justice entitle them? With respect to many of those organisms that occupy the region between the bare possession of sentience up to the onset of self-consciousness, the requirements they have for environmental resources will be essentially the same as in the case of non-sentients. But at some point, short of full self-consciousness, we have many organisms which possess as part of their life activity a very heightened awareness of much that is in their environment, including a very detailed awareness of, and need to respond to, members of their own kind. Such creatures, found among mammal groups in particular, especially primates, have developed the kind of sensitivity which it is not inappropriate to think of as a social sense.

Self-consciousness appears to exist only among the great apes, including, of course, ourselves, although there seems to be good evidence that it can be found among other mammal groups, such as dolphins and pigs. Members of these groupings appear to be able to recognize themselves in a mirror, a simple and convincing test of the possession of self-consciousness. For to do this requires the ability to have the following thought (albeit not necessarily expressible in a language): 'the creature in front of me is identical to this creature looking at it.' This necessitates possession of the concept of 'this creature doing the looking'.

The importance of these kinds of distinction for ecological justice is, of course, that, since the latter is concerned with securing for different organisms the environmental resources necessary for them to flourish 'after their kind', it is going to be important to obtain a good understanding of what such flourishing involves. This requires that human moral agents whose decisions may have an important bearing upon the capacity of non-human moral patients to survive and flourish need to possess an accurate and detailed knowledge of the welfare interests of the various species which their decisions can affect. This is the kind of knowledge which ecology and ethology are best equipped to provide, although here, as ever, it should not automatically be assumed that only the knowledge obtained by scientific investigation in the modern sense will do. Much important knowledge is likely to be possessed by traditional peoples with direct experience of many of the organisms in question. But there is undoubtedly an enormous amount still to discover about the social and psychological needs of complex animals which has a direct bearing upon their conditions of flourishing, such as the recent discovery that capuchin monkeys possess a sense of fairness and the tendency to sulk when they believe themselves to be treated unfairly (Radford 2003b).

Animals which possess high degrees of social awareness, self-consciousness and the ability to feel certain kinds of emotion, even perhaps to possess a rudimentary moral and political sense and the ability to use at least a proto-language, engage in tool use and pass on elements of culture, are creatures whose needs for environmental resources become more complex and precise. They also become increasingly individualized, possessing elements of personality and the capacity to pursue a distinctive course of development. To the extent that this happens it begins to make sense to see the claims of ecological justice as residing within individuals at least as much as among the viable populations of such beings, which are the only claimants we have so far recognized. It begins, also, to make sense to suppose that they possess welfare needs which transcend the purely material, and involve such elements as the need for the respect of others of their kind, affection and solidarity, and a rich environmental setting within which they will receive the kinds of stimulation sufficient to allay boredom.

It then becomes increasingly appropriate to conceptualize the claims of ecological justice in the case of such creatures as involving various freedoms, such as that of pursuing a distinctive course of action of their own choosing. It is this kind of argument which has led certain moral philosophers, such as Peter Singer, to argue that the great apes should be granted certain basic rights of the kind which we normally believe to be exclusively appropriate to the human case. In other words, when we are dealing with this kind of animal, at least some of the elements included in Rawls's first principle of justice become appropriate subjects of distributive justice with respect to them.

It will be useful to say a little more about the case which Singer and others have mounted for the granting to the great apes of some of the rights hitherto believed to be the exclusive domain of human beings. The Great Ape Project has been operating for the last ten years and comprises 'an idea, a book and an organization' (Kuhse 2002: 128). A *Declaration on Great Apes* formulates the basic aims of the project as follows:

We demand the extension of the community of equals to include all great apes: human beings, chimpanzees, gorillas and orang-utans. 'The community of equals' is the moral community within which we accept certain basic moral principles or rights as governing our relations with each other and enforceable at law. Among these principles or rights are the following: (1) The right to life. The lives of the members of the community of equals are to be protected. Members of the community of equals may not be killed except in very strictly defined circumstances, for example, self-defence. (2) The protection of individual liberty. Members of the community of equals are not to be arbitrarily deprived of their liberty; if they should be imprisoned without due legal process, they have the right to immediate release . . . (3) The prohibition of torture.

(Kuhse 2002: 128)

These are very strong claims, of course. They literally would permit gorillas, say, the legal right to roam at will through their preferred habitat, the legal right to habeas corpus to challenge improper imprisonment, and the right to appeal in law against their detention on the grounds of the threat they pose to themselves or other members of the community of equals. Killing of gorillas would not be permitted except in case of self-defence and other extreme situations. Presumably, then, other members of the community of equals could be charged with murder if they killed a gorilla for reasons beyond the permissible ones. Perhaps, too, a gorilla which killed a human being in self-defence (though they appear not to do so, and to be entirely vegetarian, unlike their chimpanzee cousins) could mount the successful defence of justifiable homicide.

Another possibility in such cases as the latter, of course, is for it to be held that no non-human great ape could properly be charged with a crime for the reason that they, like human children, can properly be deemed to be beyond any criminal responsibility. After all, mature human beings have to be able to meet a fundamental test of capacity – can they understand the charges which have been brought against them? – before they may be subject to a legal trial. It seems undeniable that no great ape besides human beings has the capacity to understand the idea of moral and legal restraints upon action, so no other great ape could ever properly be the subject of a legal action, even though human agents could quite properly be tried for criminal actions against them.

Of course, it is no real problem here that gorillas and the other non-human great apes would have to have their basic moral and legal rights defended by human beings, as the only fully fledged (if we allow at least some of the great apes elements of moral awareness) moral agents around. If the situation of the other great apes may be deemed to approximate that of human children, then there seems to be no objection in principle for them to be granted legal standing and legal rights, to be entitled to legal guardians to help protect and enforce those rights, and for human beings above the age of legal responsibility to be charged, found guilty and punished for any violation of those rights. After all, human children are certainly within the moral community, even if they are not expected to bear the full moral responsibility of that membership.

However, the logic of that analogy is that we expect children to be the responsibility of some adult or other, adults with the job not just of care and protection, but also with the responsibility to curb any anti-social behaviour engaged in by their charges. This suggests that, were we to grant the great apes full standing within the community of equals, that is, to treat them completely on a moral par with human beings, then we would have to allocate to them responsible guardians to ensure that their morally innocent behaviour did not harm the interests of other members of the community. In the normal human case, of course, this role falls naturally upon close relatives of the children in question – usually the biological parents. In the case of the other great apes this will not work, since all their closest relatives are equally incapable of accepting criminal responsibility.

In practice this would probably not be a great difficulty. If human beings ever get as far as including the great apes within the moral and legal framework hitherto regarded as the sole prerogative of human beings, then it will be for human jurisdictions to establish the requisite legal guardians. These would have the twofold responsibility of restraining the potentially anti-social behaviour of the great apes, and of protecting the legal rights of the great apes against the criminal actions of full moral agents. Such guardians would only be likely to do their job if they had a full commitment to the idea of the full moral standing of the great apes, but there appear to be plenty of human volunteers who would be willing to shoulder the responsibility.

However, the picture which is emerging here is of a group of non-human animals which are receiving a certain set of important protections at the cost of close human involvement in practically every aspect of their lives. This may afford them strong forms of legal protection, but would it satisfy the basic requirements of ecological justice, which up to this point has emphasized the need for organisms to have sufficient environmental resources to be able to survive and flourish 'after their kind'? In the case of the great apes it may be defensible to argue that, given their quite complex intelligences and well-developed social senses, they may not find living in close proximity to human beings who are knowledgeable of the apes' social and psychological needs to be in any way stultifying of, or harmful to, their fundamental natures. Perhaps, indeed, they may find it a rewarding aspect of their lives.

On the whole this is a matter which might well be left for further resolution in the light of studies of the great apes' psychology. It does not appear to be impossible in principle for human guardianship institutions to be sufficiently sensitive to the needs of their charges for the necessary human involvement in their lives envisaged by the Great Ape Project not to be damagingly intrusive. But it has to be noted that the presence of human beings, however discreet, is quite likely to have an effect over time on the development of the great apes themselves. This is always going to be a possibility and in many ways is entirely unobjectionable, since all living creatures co-evolve to various degrees when they share a biosphere. One thing which may lessen any concerns in this regard is that human beings are not going to go away, and thus the only real issue is not whether or not they are going to live in close proximity to the great apes, but under what terms they can be persuaded to do so.

This, of course, raises the issue of justification of all this, an issue which we have not yet addressed. The Great Ape Project is in effect selecting the great apes for special treatment, by making the members of all the great ape species full members of the human moral community. They might instead be regarded as we have so far been regarding all other members of the non-human world of organisms, namely as having moral standing and a claim in justice to certain environmental resources, but not as being equals of human beings. How is a strong claim of equality to be defended in their case?

One claim which is often made specifically with respect to the great apes is that all four kinds have diverged from a common stock very recently, in biological terms. Estimates vary, but humans and chimpanzees appear to have begun to diverge some five million years ago. The closeness of our evolutionary connection is manifested in the similarity of our DNA. Again different estimates occur of the common percentage in different writings, but the usual figure is in the region of 98 per cent to 99 per cent (Diamond 1992: 18). Jared Diamond famously proposed that the closeness of genetic fit between human beings and the two recognized species of chimpanzee justifies regarding all three species as members of the same genus (ibid.: 21). At the very least, no other animals appear to be more closely related to human beings than the great apes.

The significance of these facts in the course of arguments for the admission of the great apes as moral equals of human beings is that they appear to warrant a strong emphasis on the continuity between human beings and other living entities, rather than supporting the claim of a bottomless and unbridgeable chasm between human beings and other species. However, they do not in themselves warrant a conclusion of moral equality between members of the great ape species of the kind which the Great Ape Project puts forward, for similarity is not identity, and the remaining differences between humans and great apes might still be held to warrant a clear moral distinction between them. However, bringing the facts of our recent shared evolutionary past to the forefront of the debate performs the admirable service of rendering more initially plausible the hypothesis of moral equality. For if we are so similar in terms of our DNA it seems unreasonable to rule out without further examination the claim that we have other, morally more relevant, characteristics in common which would justify a strong claim of moral equality between us.

In fact the case for the Great Ape Project does focus on the characteristic that is arguably absolutely central to possession of full moral status, which is the possession of personhood. The argument is that there are sufficient similarities between the mental traits of the great apes and human beings to warrant the claim that great apes are persons just as much as are human beings. Of course, the evidence does not show that even the great apes which are closest to us, such as the pygmy chimpanzees, or bonobos, are just like mature human persons in every significant respect. Rather, the argument goes, it can be shown that great apes possess sufficient of the personhood-conferring capacities to a sufficient degree to justify their classification as persons, with the implications for their full moral equality with mature human beings. What are these traits? We have already noted the capacity of great apes for self-awareness. They have also been credited with linguistic capacities, including the ability deliberately to deceive others, involving the ability to put oneself imaginatively into the position of others in order to work out how something is going to look from their point of view (see, for example, Van Lawick-Goodall 1971: 102–4). Some appear to have the capacity for guilt, shame, or anger at being found out. Some have recognizably political forms of interaction, for example making and breaking alliances in order to ascend the slippery pole of chimpanzee society (see De Waal 1998). Tool use and the transmission of rudimentary traditions seem to be present too. It is possible to object to these claims on the basis that they involve anthropomorphism, operant conditioning or just romanticism on the part of human observers. However, from the point of view of Darwinian theory there is absolutely no reason to set oneself resolutely against the possibility of elements of personhood on other species. For one thing we do now have evidence that up to about 40,000 years ago we did share this planet with another species which probably had as full a claim to personhood as we do ourselves, namely the Neanderthalers. These now appear to have been a separate species from *Homo sapiens* rather than a subspecies (see *Neanderthalers and Modern Humans: a regional guide* 2004). But even if new data turns up which leads to this view being abandoned, the point is that there is absolutely no reason why we should expect that only our own species has ever attained (or can ever attain) at least some of the traits of personhood.

This latter point serves to emphasize the other crucial element in the argument, namely that personhood is not an all-or-nothing trait, like the possession of an indivisible soul. If nowhere else we can see this within our own species. We acquire personhood a step at a time as we mature. We may lose or never acquire elements of personhood – a sense of self, say, or of the capacity to make sense of others – as the result of malformation, accident, ageing and so forth, while retaining others.

If we also accept, as there appears to be good reason to do, that it is the attainment of personhood to some degree which confers full moral status on an organism (and, once again, it is arguable that there can be no disembodied persons, although that is another argument not directly relevant to the matter in hand), then the case for the Great Ape Project begins to look in quite good shape. For it is now possible to argue that (1) personhood is a complex phenomenon which admits of degrees; (2) the possession of some degree of personhood by an organism is sufficient to establish its full moral status; (3) the great apes possess personhood to a sufficient degree; (4) hence the great apes possess full moral status.

None of these claims will be without its detractors. However, there is at least a strong prima facie case for the admission of the great apes to the community of justice as full equals with human beings, on the basis of a comparison of their status with that of human children prior to the age of full moral and legal responsibility. As already noted, the practical implications of this view do not appear to be insuperable, although getting the acceptance of sufficient human beings in a position to implement it is another matter. As we will be discovering in Chapter 12, there currently exists a United Nations initiative, known as GRASP – the Great Apes Survival Project – that superficially resembles the Great Ape Project.

However, as we will see, this project, which has the entirely laudable aim of trying to save the great apes from extinction, is founded on a purely anthropocentric basis, namely that there are various important human interests which may be served by saving the great apes. It will, however, be worth examining that project more fully from the point of view of ecological justice in due course.

To cast the matter in terms of ecological justice, then, we can ask to what environmental resources the great apes are entitled if we accept that they are also full members of the community of justice. Here is what Peter Singer and Paola Cavalieri say on the matter:

> guardians will be needed to protect the newly sanctioned rights to life, liberty and freedom from torture of the other great apes . . . even the idea of protected, independent territories where nonhuman great apes could regain the dignity of autonomous life is not without precedent, as the existence of human regions in need of external protection, the United Nations Trust Territories, shows.
>
> (Kuhse 2002: 140)

This seems to be arguing for the great apes to be ceded territories of their own – quasi-states, perhaps. Many pressing questions immediately arise, of course. Will human beings, except for the guardians and their representatives, be kept out from such territories? Or will human beings be permitted to live within them, making use, in some way which respects the rights of the great apes, of the environmental resources of the territory? If the former, how is it to be achieved, unless the territories are so impoverished of resources that no human beings will seek to gain access to them? If the latter, how will the human beings who reside within the trust territories be governed? Will they be allowed to have their own democratic government? Or will rule be by the (UN?) appointed guardians? Either way, how do rights for humans relate to the rights of the great apes?

These are difficult issues. But it has to be said that there is no way of trying to save the great apes which is not fraught with difficulties. Whether we try to save them and allow them to thrive because we think only that there are important human advantages to be pursued by doing so, or because we believe that ecological justice requires it, or because we believe that they are the moral equals of human beings, there will need to be great deal of expenditure of time, money and effort to achieve that goal. Even if we argue that the great apes should produce some of their own income, via tourist revenues for the right to visit them in their protected areas, that is unlikely to cover all the necessary costs. And, of course, it is a presupposition of moral personhood that you possess the rights of a person whatever interest others take in you. If no human being ever again had any interest in paying money to go to visit the great apes in their protected territories they would still have the right to go on living in them.

Other issues which also become pressing concern what rights if any the great apes would have to move beyond their allotted territories. Presumably this would have to be restricted and probably prevented entirely, for their own sakes as well as for the sakes of other creatures, including human beings. What if their numbers increase to the point that the existing territories are no longer large enough to support them? Is it permissible to kill some of them – for the good of the species as a whole? This would start to sit rather uneasily with the idea that they have full moral equality with human beings. We certainly do not think that we can cull human beings in order to regulate population sizes. Would some form of enforced contraception be justifiable?

In the face of such difficulties it will certainly be difficult to do complete justice to the idea that the great apes are fully equal members of the community of justice with human beings. What should the attitude be of proponents of ecological justice to the Great Ape Project in the light of these difficulties? The project puts forward the strongest possible moral case for the full equality of human beings and other organisms. It nevertheless rests a large amount on the idea that the great apes possess the traits of personhood to a sufficient degree to justify that claim of equality. There is no obvious logical impropriety in denying that they do possess traits of personhood to a sufficient degree to warrant that judgement, even if there is no real room any longer for denying that the great apes possess some of the elements of personhood in some degree. Thus, supporters of ecological justice clearly have no logical commitment to the Great Ape Project, although there is nothing within that project to preclude its acceptance as a key component in the project of ecological justice either. This also implies that supporters of ecological justice may without impropriety judge instead that the great apes do not possess sufficient degrees of personhood to justify the attribution to them of full moral equality with human beings, but that they nevertheless are entitled in justice to their fair share of environmental resources, and, given the proven facts of their individuality, that this claim resides in them as individuals rather than as populations. Given the difficulties which both approaches are likely to have with securing the survival of the great apes in practice, this difference might best be put aside until the moment is reached when some matter of practical import turns on its resolution.

Even if we are contemplating the claims in justice of less individualized creatures, such as many of the other social mammals, which possess social awareness but not individual self-awareness, it is arguable that their environmental needs must be taken to encompass the resources, such as the opportunity to acquire and maintain territories, which enable that social awareness, so essential to their kind and its way of life, to be expressed.

Thus, as we explore the realm of the sentient and more-than-sentient organisms we discover that the conditions for survival and flourishing of their kind become, unsurprisingly, increasingly complex and thereby pose an increasing challenge to the exponents of ecological justice. The competition for environmental resources between human beings and other animal species becomes even more difficult to resolve when those other species can be shown to have environmental needs that sit uneasily with extensive human use of those same resources. This is not, of course, a problem which exclusively applies to complex more-than-sentient organisms. Many non-sentients, such as some plants, may have developed in such a way that they occupy highly specific niches in the environment, such as island ecosystems, to which human beings may pose grave threats. But the requirement to maintain in existence small and rare habitats is arguably less of a challenge than that of maintaining extensive habitats, such as savannah and rainforest, within which alone certain complex social animals can follow the form of life appropriate to their kind. This will apply even more acutely in the case of animals, such as lions and eagles, whose life-ways involve predation of complex social animals.

It is, of course, with respect to these kinds of animal that the standard solution of creating a delimited area within which they may pursue their life-way is most appealing. Many of these have been created, although not for reasons of ecological justice. They have been created, as in the USA, for the benefit of the citizens of the state concerned – for their recreational, cultural and aesthetic benefit – or for the economic benefits accruing from the money which human beings are prepared to pay to get access to such areas for the benefits just mentioned. What difference, if any, would be made to these park areas if they were to be established as a solution to the requirements of ecological justice?

The first is that their creation would be morally, rather than prudentially, required, if they proved to be the only viable solution to the problem of securing the appropriate environmental resources for the kinds of animal we have been considering. This element of moral compulsion raises serious issues of responsibility, something which has not yet been directly dealt with. It is a topic to which it will be necessary to return in the next two chapters when we consider the institutional arrangements required by ecological justice. Suffice it to say at this point that ecological justice is ultimately a joint responsiblity of all the moral agents on the planet, who must be prepared to offer support to those among them whose occupation of specific parts of the planet makes them directly able to influence in those areas the environmental conditions upon which the survival and flourishing of various non-human organisms depend.

However, it is no longer as obvious as it once appeared to be that the setting aside of protected areas in order to enable endangered species to survive and flourish, whether for reasons of ecological justice or for the anthropocentric reasons mentioned above, is really viable. Again this is a matter to which we will return in the next two chapters.

At this juncture, however, it is appropriate to address an issue which has not yet been examined, and which is implicit in the very idea of ecological justice. This is the point that ecological justice is supposed to apply to all organisms which occupy a biosphere. This, in the case of planet earth, encompasses the human, as well as the non-human. Hence, it is immediately implied that human beings also have a claim in justice to a share of the planet's environmental resources sufficient to enable them to survive and flourish. Human beings are, uniquely as it has turned out, though contingently, the only organisms which are autonomous moral agents. They are the only beings capable of recognizing the existence of ecological justice and of seeking to secure it. How are they to determine what ecological justice entitles them to claim?

Of course, what is immediately apparent is that ecological justice for humans is essentially identical to at least a large part of the topic of distributive justice among humans as traditionally conceived. The issue of how human beings are to determine what each may properly claim as his or her fair share of environmental resources is certainly partly about survival, but it is in the area of 'flourishing' that serious complications arise. For here it is a matter of protracted and ongoing dispute what forms of life for human beings may properly be encompassed within the concept of flourishing. Liberalism tries to leave the matter largely for individual judgement, limited only by the consideration that human individuals, in their pursuit of their view of the good life, so act as to leave others the opportunity to pursue their, probably different, view. Other political views have been more overtly perfectionist, and in many cases objectionably authoritarian in their approach.

We are here back in the territory of the last chapter, except that we have arrived at it by considering how the requirements of ecological justice become more complex as the psychology of the organisms we are considering also become more complex. What this shows, I suggest, is that ecological justice as this book has characterized it is the more fundamental starting point for the development of ideas about distributive justice than the traditional focus upon human groups and the conditions for them to engage in reciprocal recognition. It is unsurprising in retrospect that human moral agents should have begun at this place, since it is the point of maximum practical concern to them as history develops. But we can now also see retrospectively that the conception of distributive justice which emerges on this basis focuses on only a part, albeit the most complicated and, for humans, the most immediately interesting part, of distributive justice.

When that understandable oversimplification is remedied, however, we once again see the need to modify the traditional liberal view of the pursuit of human flourishing in a crucial respect. For the justifiable restrictions upon the pursuit by human individuals of their own vision of flourishing need to contain an extra element, namely the requirement to take account of the conditions of survival and flourishing of non-human organisms.

Recapitulation and further points

What the last two chapters have sought to show is that it is possible to outline the key elements of what ecological justice requires. Once we accept that all living organisms, sentient or non-sentient, possess moral considerability, by virtue of their possession of morally significant welfare interests, then it becomes a matter of distributive justice that they receive their fair share of the environmental resources needed to meet those interests from those moral agents who are in a position to affect the distribution of those resources for good or ill. We argued that in the case of non-sentient organisms the claim to the resources resides in the naturally viable populations of the species in question, rather than in individual members of the species or the species as a whole. Lacking a psychology, non-sentient organisms are not individualized. What is important from the point of view of their claim to resources is that enough is available to each population to keep it viable and its individuals surviving and flourishing within the rather narrow range of that concept as applied to them.

Sentient organisms have rather more readily been granted moral considerability than have non-sentients, but for many of them too the same arguments of ecological justice apply as do to non-sentients. Although possessing sentience, many such organisms are not highly individualized, and their claim in justice to fair shares of environmental resources applies to viable populations rather than to individuals or the species as a whole. As sentience becomes a more elaborate phenomenon, and as individual members of sentient species acquire increasingly refined forms of awareness of their fellows and their surroundings, the claims of individuals start to become more prominent. To put the matter more concretely, it starts to become increasingly difficult to justify the killing of an individual member of the species. Killing increasingly needs to take account of the fact that the creature threatened with death is distinct from its fellows and has an identifiably different life to live from theirs. The emergence of self-awareness and elements of personhood take this process even further, until, as we have just seen, it starts to make sense to attribute to at least some organisms equality of moral status with human persons.

Human persons themselves also, like all other organisms, have claims in justice to environmental resources. These claims are vested in individuals and held against their fellows. The working out of what claims they have and how they are to be met is embedded in the topic of distributive justice as traditionally conceived. But the decisions concerning ecological justice are inherently directed towards all beings with a justifiable claim, which is all living entities that share the biosphere with human beings.

The crucial issue of how to resolve clashes of interest between different organisms has partly been touched on. The matter has already been examined in an earlier publication, and so what follows is in effect a brief summary of that discussion (Baxter 1999: 79–83; 151–4).

Clashes of interest between different organisms are to be decided by means of a distinction between basic welfare interests and non-basic ones, and between the different moral weights which can properly be attached to different kinds of organism. Basic welfare interests of less weighty organisms (populations), such as that of surviving and flourishing, will trump the non-basic welfare interests of more weighty ones. When equally basic interests of organisms with different weights clash, then the interests of the more weighty organism trump those of the less weighty.

When a species of organism is facing extinction, then, however lowly it is, its weighting receives an automatic boost. It will only be possible to trump the claims in justice of such an organism if the basic interests of the most weighty moral being (a human individual's life being saved) can be shown to be directly opposed to the basic welfare interest of the endangered organism. However, even in some cases, where the human being's interests can be argued to be not very great (perhaps they are in terminal condition), it would be justifiable to allow the claims of the endangered less weighty species to trump those of the human being.

It might be thought that all these claims suffer, at the very least, from an incapacitating vagueness. As they stand that is probably true. But the crucial issue is whether they can be made operational by suitable further debate. It is not at all obvious that this will be impossible. All theories of justice have to enunciate general principles and leave individual jurisdictions to put some legal beef on them and find out ways of resolving the complexities involved. It is hard to see that this is avoidable. Even if we agree that social and economic inequalities are to be so established that they work out to the greatest advantage of the least well off, or that principles of distribution should be determined by the social meanings of goods produced by a specific society, or that people are entitled to holdings which they have acquired without violating the rights of others, these operative phrases all need to be pinned down and refined before they can be applied.

What is basically important about a theory of justice, ecological or otherwise, is its fundamental tenor or shape – what it is seeking to achieve and on what basis. Ecological justice seeks to pursue what its exponents regard as the basic insight that all living entities which share in, and jointly sustain by their activities, a single biosphere are prima facie entitled to claim against each other, or to have claimed on their behalf, a fair share of the resources which they jointly make available within that biosphere, subject to the right to self-protection. If this is a basically sound position, then the difficulties which inevitably attend its operationalization will properly be regarded as challenges rather than death-blows.

However, in the next two chapters we will need to say something more detailed about how ecological justice may be put into practice in the actual world, and it is to this task that we now turn.

Part IV

Institutional arrangements for ecological justice

11 Institutional arrangements within states

The concerns of this book so far have been focused almost exclusively on issues of moral and political theory. We have sought to vindicate a universalist approach to moral theory in general and then to apply this approach to the generation of a theory of justice in the distribution of environmental resources among all species of organism on this planet. This theory of ecological justice has aimed to produce a case for encompassing all forms of living entity – from the 'merely' living to those creatures that appear to have achieved some of the elements of personhood which are most fully exemplified in the case of one species, *Homo sapiens*.

In the course of this discussion there have emerged various suggestions and implications for the ideal form of institutional arrangement which should be aimed at in order to pursue the goal of ecological justice. It will be useful to have before us a brief outline of the ideal theory before we consider what the possibilities are in the real world for the pursuit of ecological justice.

The ecological version of justice as impartiality argued for the inclusion of all organisms within the community of justice. This implies that, where a group of moral agents are establishing or revising a political system, and are seeking to find a set of basic political arrangements which all can agree to live under, by searching for those which are 'not reasonable to reject and so fair', then they are bound to consider the welfare interests of all the organisms which their activities can affect for good or ill.

We imagined a set of human proxies seeking to defend the weighted interests of the non-human within the constitutional design of the polity. As we have previously noted, this implies the reasonableness of a constitutional provision for the ongoing protection of the consideration of those interests. At a minimum this will require the constitutionally entrenched provision of some institution with the responsibility to ensure that the welfare interests of all the relevant non-human 'inarticulates' are presented within the polity's decision-making bodies, at central and local levels. This in turn will be a matter of enforcing with the polity's legal apparatus the application of a set of evaluative criteria for proposed courses of action of the kind which we elicited from Wetlesen's list in chapter 9.

The point of the constitutional provision, it will be recollected from our discussion of Bell's version of Rawlsian political liberalism, is to make ecological justice difficult to avoid, even though no force on earth can oblige a society to retain such a provision if it has a mind to do away with it, or to pay attention to it if it keeps it. No guarantees are available here – the best we can do is to try to tip the balance in favour of ecological justice as far as we can. As we noted in discussing Bell's arguments, it does not follow at all that what is being proposed here is tantamount to a form of authoritarianism or oligarchy.

It follows that the ecologically just society will make constitutionally protected arrangements to give some human agents the responsibility to ensure that the welfare interests of non-humans are articulated and taken full account of in the business of legislation and policy-making in all public forums, and within private spheres of action such as business, private associations and family life. I have elsewhere put the case for an elected, but partially insulated, guardianship body within national polities to do this job (Baxter 1999: 125). Dobson has argued for proxy voting for elected representatives standing for the non-standard constituencies – future generations, contemporary aliens and the non-human (Dobson 1996). There may be other possibilities. It is difficult to determine the correct view to take from the standpoint of ideal theory. What is important is that the case is made for some workable form of constitutional protection for the welfare interests of the non-human within human polities.

However, whatever structures are devised to ensure that the public decision-making procedures of ecologically just societies take the interests of the non-human into account, institutions require human moral agents to work them, and they must have a grasp of the point of those institutions and a commitment to making them work. These considerations point to the need to ensure that appropriate education occurs within the ecologically just society of as many citizens as possible with respect to the values and arguments which underpin the institutions of ecological justice. This will undoubtedly involve societies making a commitment to a controversial substantive moral position, as we saw in the discussion of Barry's arguments. But we also saw that, even if we can envisage our polities remaining largely neutral between rival conceptions of the good life, we cannot envisage them, as matters currently stand, as being neutral between all substantive moral positions. Hence the lack of complete neutrality is not a specific weakness of the ecologically just society.

Ecological justice, however, like the ideology of ecologism of which it is a constituent part, is inherently a global phenomenon. That is, it cannot rest content with being restricted within particular polities. If its arguments are correct then ecological justice must be secured throughout the biosphere. In this respect it differs from other universal doctrines, such as traditional liberalism, socialism, Christianity, Marxism and so on. The latter claim universal validity for themselves, but they can, in principle, rest content with embodiment within individual polities. It makes sense to speak of liberalism, Christianity, Marxism and so forth 'within one country', although practicalities may mean that the ideology has to be spread fairly far before it can be regarded as secure within any one country.

But a proponent of ecological justice cannot rest content with a single ecologically just society. The aim of such a proponent is ecological justice, not simply individual ecologically just societies. It has an all-or-nothing quality as a moral commitment, and peaceful coexistence with non- or even anti-ecologically just societies is not a meaningful option. If the world contained only one society within which non-human species received adequate protection of their welfare interests, and in every other they were subjected to large-scale extinctions, then the single ecologically just society would have been largely futile. Even if the society in question were a species-rich hotspot, this would at most lessen the futility, not completely remove it.

By contrast, liberals, Marxists and Christians might also thirst after a world made over in terms of their favoured prescriptions, but they would not regard the single liberal, Marxist or Christian society as futile. Rather they would view it as having supreme value even in the absence of the universal acceptance of the favoured doctrine.

The point of these comments for present purposes is that ecological justice commits its proponents to seeking effective global institutions for the securing of such justice. This might mean a commitment to a single world government embodying the equivalent of the constitutional protections which we argued for in connection with the single polity. But there are strong arguments of principle against such a form of government, such as the possibility it creates of a global tyranny. However such issues are resolved in theory, ecological justice requires that serious thought be given to the securing of ecological justice at the global level. This is matter made especially difficult by the fact, just alluded to, that different regions of the planet are more important than others in terms of their contribution to the existence of large numbers of species. Hence there will need to be a concentration of effort in certain areas, and the obvious issue is how to ensure that the costs of such an effort are themselves allocated justly among human beings.

Having sketched the rough outline of ideal institutions for securing ecological justice, we will in the remainder of this chapter and the next turn to the practical matter of how such a goal of ecological justice may be secured by means of suitable institutional arrangements within and between states. This formulation of the issues for discussion may immediately be criticized as begging a rather large question. For some will strongly argue that it is the existence of the state, especially in its modern form, as the primary mode of human political organization that poses the greatest obstacle to the successful achievement of any environmental goal, especially a goal as ambitious as that of ecological justice. Hence, on this view, while states continue to exist, no institutional arrangements will be able to do a satisfactory job of securing important environmental goals, as opposed to trivial, placatory gestures towards them.

The critique of the oppressive power of the state has, of course, for long been the hallmark of anarchist theories, and there is undoubtedly a strong anarchist strain running through much contemporary environmental moral and political theorizing. Murray Bookchin, famously, argues for the political reconstitution of the world as a federation of municipalities organized along participatory, non-hierarchical lines (Bookchin 1995). We have seen the explicitly anarchistic approach of Mick Smith, whose 'ethics of place' involves a thorough-going critique of the modern state and its commitment to a destructive, oppressive, rationalist universalism (Smith 2001). Alan Carter has recently offered another wholesale critique of the state as locking human beings into an environmentally vicious cycle which can be broken only by dismantling the structures of the state themselves (Carter 1999). The 'global ecology' approach associated with Wolfgang Sachs also emphasizes the need to protect the remaining commons of the world, governed by the traditional commune system which may still be found within the states of the global South, from the depredations of authoritarian, centralizing states (Sachs 1993). The loose movement of groups opposed to 'globalization' and its adverse effects on poverty-stricken humans and the environment exemplify the same kind of critique in the world of practical political activity. This list might be extended in various ways, but enough has been said to suggest that there is currently a deep-seated disquiet about the state, particularly in its liberal-capitalist form, on the part of many environmentally concerned people.

The issues surrounding the role of the state raised by the writers mentioned in the last paragraph, and others, deserve careful consideration. There is much in their critiques with which one may wholeheartedly concur, especially the pernicious effects of the entanglement of the state with the military–industrial complex in the liberal capitalist states of the global North.

For the purpose of this book, however, analyses of the threat posed by the state which require its wholesale transformation into some other political system will be set to one side. The aim of securing ecological justice is already, many would argue, a rather strongly utopian one. It may be that this utopian goal will itself only be achievable if another utopian solution is found to the existence of the state, a tall enough order in itself. However, the two revolutions here being considered are logically independent of each other. At most the establishment of anarchistic, non-hierarchical, democratic forms of human political organization is a necessary condition of the achievement of ecological justice, but it is clearly not sufficient. Human beings may remain strongly anthropocentric in their moral outlook, even if they are organized in non-hierarchical systems (Bookchin, for example, has little patience with non-anthropocentric moral positions). If you are convinced by the case for ecological justice and, for the reason just given, not inclined to pin all your hopes for its attainment on the prior removal of the state, then you will be strongly inclined to examine the possibilities for attaining ecological justice within the world as it currently is.

Hence, for the remainder of the book it will be taken for granted that for the foreseeable future the state will remain the primary mode of political organization across the globe, with all its problems and drawbacks – and, of course, its advantages too. The justification of this is that the case for ecological justice, if it is sound, requires us to do our best within current arrangements to achieve its goals. Of course, 'ought' implies 'can', and if we cannot achieve ecological justice within our existing political systems then we will have no reason to attempt to do so, and our aims as proponents of ecological justice will indeed have to turn to securing the political prerequisites for our moral goals. However, there are some reasons for supposing that, even within current political arrangements, something effective may be done in terms of setting up institutions for the pursuit of ecological justice.

The first such reason is that the world currently contains, both within and between states, a plethora of institutional arrangements for the pursuit of a whole variety of environmental ends. Of course, there are many reasons to view this fact with a cautious, or sceptical, eye. Not every state contains such institutions, or many of them. Even where they exist they may not work well, or even at all. Not every state which purports to support international environmental arrangements does so in practice, and some may even work to undermine those which they nominally support. Even where institutions do exist and do work reasonably well, within and between states, their continued existence, or effectiveness, cannot be taken for granted. The normal working of democratic politics may lead to changes of policy direction within states which have been environmentally progressive, as political personnel changes in the course of the electoral cycle. The adverse effect of the operations of environmental institutions on the important interests of many groups will give the members of those groups a continued interest in undermining and abolishing institutions which have been successfully introduced.

However, even in the face of such difficulties, the continued existence and at least intermittent success of some institutions aiming at protecting some aspect of the environment mean that it is not a completely forlorn hope that the defenders of ecological justice will at least have as much success in achieving their aims as have other environmentally concerned groups in achieving theirs. Arguably, too, where a goal rests on a moral justification it is incumbent on those who pursue the goal not to give up its pursuit in the face of difficulties and disappointments.

The second reason for a more positive assessment, from the point of view of ecological justice, of the institutional arrangements attainable within a world of states is that quite a number of the environmental institutions already established, however problematically, are devoted to the protection of endangered species and their habitats. The oldest non-governmental organizations (NGOs) devoted to environmental causes are usually conservation or preservation bodies, such as the Royal Society for the Protection of Birds and the Sierra Club, which have arisen within states, usually in the late nineteenth century, in response to threats to species and habitats within the borders of the states in question. The governments of many states have responded to the political prompting of such groups by implementing protective legislation and establishing bureaucratic organizations to oversee its implementation.

At the international level, too, institutional arrangements have been created to protect endangered species across state boundaries, such as the Convention on International Trade in Endangered Species (CITES), and sometimes whole environments and ecosystems have been the subject of international institutional arrangements, such as the Antarctic Treaty. In the main, such efforts have received their justification in purely anthropocentric terms. What is held to justify the time, effort and expense (in terms of direct and opportunity costs) involved in the setting up and servicing of such institutions is that there is some important human interest or other which is at stake – whether non-material, such as cultural, spiritual, recreational, scientific or aesthetic interests, or material, such as the preservation of valuable food stocks, in the case of fisheries protection.

These facts are of interest to the defender of ecological justice because all these institutions are operating with a goal similar to that of ecological justice, even if not for the same reason. One possibility that this suggests is that at least some of these institutions may be induced to alter the rationale for their activities from one which is purely anthropocentric to that offered by the theory of ecological justice. This will at the minimum give an opening for the further development of institutional arrangements in the direction desired by the proponents of ecological justice. Another important dimension which requires exploration is the discovery of what lessons may be learned from the experience of conservation institutions by those seeking to fashion institutions of ecological justice.

It is important to note the drawbacks and deficiencies of such institutions, too, from the point of view of ecological justice. The first is that their concerns are inevitably partial. This partiality manifests itself in at least two directions. Firstly, the concern of an institution typically is to protect some subgroup of organisms which fall into a specific category – birds, or sea mammals, or creatures which are traded, or all species within a delimited area. Organisms outside of these categories may not receive any protection, except indirectly as a result of their presence in habitats which are preserved as part of the effort of protecting the target group. Secondly, the protection afforded is for a specific reason which does not cover all organisms, such as the cultural value of a species to a given society (the bald eagle), or its economic value (the cod), or its religious significance. Some of these reasons may be powerful within a given society and afford the species in question a great deal of protection. But these reasons are all inherently contingent. If a religion declines, an alternative source of food is found, or people lose aesthetic or recreational interest in a set of organisms or habitats, then these institutional arrangements are put in jeopardy.

The reasons offered by the theory of ecological justice for protecting non-human organisms suffer from none of these drawbacks. Other creatures, even those holding no specific interest for human beings, are given the right to the environmental resources needed for their survival in virtue of their moral standing. The latter, once established, is not subject to the specific vagaries of human culture just noticed. Of course, nothing can guarantee that any moral case, once accepted, will always remain so. But this is unavoidable. Any theory of justice is always subject to one ineradicable source of change, namely that in human understandings of morality.

The drawbacks just noted mean that existing institutions of species protection are at best starting points for ecological justice. This is not to say that the organization of ecological justice may not usefully employ various subdivisions of concern to make the effort of protection manageable. For example, within a given state the key institutions of ecological justice may focus on species or groups of them which are under particular threat within the territory of the state. But the institutions must always at least be alive to the claims of various species as environmental factors change, whether for reasons of human activity or as a result of developments emerging from non-human nature itself.

To return to reasons drawn from the actual state of the world for supposing that institutions of ecological justice are not a forlorn hope, we need to note that in some states there are the beginnings of constitutional protection for the non-human. The discussion of justice as impartiality revealed the importance of establishing within all states the constitutionally embedded recognition of the moral standing of the non-human and of its claims to a fair share of environmental resources. In Germany this has recently been recognized in terms of constitutional protection for animals. This has been motivated more by animal welfare concerns than by environmental ones, but its significance is that it gives constitutional recognition to the moral standing of non-human creatures. Of course, animal welfare legislation is in itself not a new thing. But the constitutional embedding of animal welfare marks a new departure. One might say that it makes such animals part of a human polity for the first time. This is a change in their status, rather than a change in the kinds of claim they have long been recognized as having upon us.

The interesting point which emerges from this survey of existing institutional arrangements for the protection of various aspects of the non-human world is that many of the ingredients for satisfactory institutions of ecological justice are already present in many states. Many states, as we have seen, have institutions designed to protect species from various threats which human beings make to their continued existence, and have also agreed to the creation of international institutions to do likewise. But these institutions are underpinned by anthropocentric arguments which do not recognize the moral claim that non-human species have upon us. Many states also have animal welfare institutions designed to protect animals from certain types of attack by humans, intentionally or otherwise, on their welfare. These do recognize that at least some non-humans do have a moral claim upon us, but their focus is usually upon the sufferings of individual animals deemed to have the capacity to experience suffering. Provided no suffering is involved, such institutions put no limit upon the depredations made by humans upon whole species. Extermination of species, provided it is carried out with no, or very little, suffering, does not fall foul of the penalties properly exactable by such institutions.

What proponents of ecological justice need to do, therefore, is to marry together (1) the recognition of moral status of the non-human which is embodied in welfare legislation and (2) the species-oriented approach of conservation legislation, and seek to develop the whole package in the direction of constitutional embedding exemplified in the recent German case. It is a virtue of focusing our discussion upon actually existing institutions in the problematic world of states, rather than elaborating a utopian vision of ecological justice in a world totally different from our current one, that we can keep continually before us the significant developments which have actually taken place. This strengthens the case for ecological justice against those sceptical minds which claim the whole idea is hopelessly utopian and alerts us to those problems which have arisen in the course of developing the actual institutions which we see before us, thereby providing important tactical and strategic lessons.

For reasons of space it will not be possible to examine all the relevant institutional phenomena, such as:

- endangered species legislation (wildlife protection acts)
- habitat protection legislation
- biodiversity action plans at local and national levels
- legislation for protection of animal welfare in food production/agriculture/ scientific research
- protection of economically-important species (e.g. fisheries)
- legislation against blood sports and recreational attacks on animal welfare (badger-baiting, etc)
- constitutional provisions for animal welfare.

But we can justifiably make the following points from the existence of these kinds of legislation and enforcement procedures:

1   It is possible to legislate in all these areas, and to have the legislation success-fully enforced.
2   Such legislation is capable of receiving widespread support within the societies within which it operates.
3   Such legislation involves acceptance of restrictions, some of them severe, upon human beings' pursuit of their own interests.
4   It is possible for human beings to 'speak for the non-human' in a non-arbitrary and motivated way, basing claims upon convincing evidence (about, say, what non-human preferences actually are).
5   The existence of strong feelings, controversy over 'the facts', difficult issues of balancing interests, competing value-judgements and so on do not need to prevent effective action being taken to conserve endangered species and protect non-human welfare.
6   It is possible, therefore, to countenance the constitutional embedding of the rights central to the claim for ecological justice within states and to envisage appropriate institutional arrangements for the protection of such rights.

It is important to begin the examination of appropriate institutions for ecological justice with the intra-state case because, as we will discover when we turn to the issue of international institutions in the next chapter, the crucial ingredient in the successful creation of effective international institutions is the existence within at least certain key states of an effective intra-state regime of environmental protec-tion, involving NGOs, state bureaucracies and other actors. States have govern-ments, and not just systems of governance. Governments, whether democratic or autocratic, have at least in principle the capacity to legislate and enforce their legislation. The state may be problematic in all the ways referred to above, but this legislative function remains crucial and cannot be neglected by the proponents of ecological justice. As we will also see, some states have vibrant civil societies, and it will turn out to be important from the point of view of environmental protection in general, and ecological justice in particular, that there be such civil societies. Finally, pluralism and democracy will be crucial also for the successful pursuit and institutionalization of ecological justice. States which possess vibrant civil societies, democracy and pluralism are essentially liberal states. Ecologically just states will admit the non-human into the sphere of justice, which will involve the significant alterations argued for in previous chapters to traditional liberal conceptions of the role of the state. Thus a state which does all these things will in effect be an ecologically just liberal state.

Any discussion of the appropriate institutional arrangements within states to secure ecological justice will do well to consider some of the difficulties and challenges which have been revealed by existing arrangements for the protection of endangered species and their habitats. Many states have set up areas within their national territories which are intended to provide protection for endangered species of organisms from all taxa. They have established official bodies to oversee the financing and management of such areas and trained and employed conservation managers versed in relevant scientific disciplines to put into effect the desired protection. Some successes have undoubtedly been achieved, but many difficulties have emerged.

Some are to do with the changing scientific basis on which conservation management rests. Some are to do with the development of disconnections, for institutional reasons, between conservation managers and their colleagues in pure science. Some concern the faultiness of the ecological understanding operating in the minds of decision-makers, such as politicians and bureaucrats, when the conservation practices were initiated. Some concern the difficulties of balancing the variety of goals established by legislatures for conservation managers to pursue, goals which emerge as part of the inevitable political bargaining necessary to establish sufficient consensus to enact the legislation in the first place. Some concern the lack of real support for the conservation effort, in terms of money, personnel and other prerequisites, on the part of the governments of states and their electorates.

It will be useful, therefore, to summarize briefly some of the lessons which the actual practice of managing human activities (for that is what is inevitably involved) to secure the protection of endangered species has revealed. Although this effort has not been engaged in for reasons of ecological justice, the latter has the same goals as standard, anthropocentrically based species protection efforts, and its proponents will do well to ponder the institutional lessons which may be learned from those efforts.

Lessons from the practice of conservation[1]

It will be useful to distinguish the following focuses in the area of conservation management:

- ecological theory and conservation management
- resources for successful conservation management
- the role of other stakeholders
- the role of the media and education
- the role of governments.

Ecological theory and conservation management

Once upon a time the science of ecology seemed to justify a fairly straightforward view of how to go about establishing areas of the globe to be devoted to the conservation of species and habitats. The idea was to locate areas of pristine wilderness containing the target species and establish a cordon around them. Within those cordoned-off areas, the natural ecosystems would reach a stable climax form, and then the task was simply to protect the areas in question from unnatural influences, such as fires and adverse human impacts. The ecological theory which underpins this view of conservation is sometimes referred to as the Clementsian view, after the work of the ecologist Frederick E. Clements. In the last thirty years or so ecology has changed its paradigm from one which sees the 'balance of nature' as inhering in such stable climax formations to a view of ecosystem development which posits no final, natural, point of stasis. Instead, ecosystems are understood as possessing a dynamic patch structure, in which developments within one area are affected by influences from adjacent areas, and the overall pattern of development will differ from case to case, depending upon the specific configuration of the patches and their changing nature (Wiens 1997).

If there are no stable climaxes in nature, if fires and other chance events are inherent in natural systems and if, further, human activity has played an essential part in the creation of patch features important for the existence of certain kinds of habitat and their associated species, then the idea of what is involved in conservation has to be drastically altered. It cannot involve just fencing off an area and keeping it stable, pristine and free from most human activity. What happens within it will be a function of a whole variety of interacting factors, such as its specific internal configuration of patches, chance events, influences from the immediately adjacent areas, including human activity, and influences from even further afield, such as global climate change. As Christensen explains (Christensen 1997: 167–86), the activity of managing areas of the globe for conservation purposes involves intervening in inherently complex and unpredictable situations.

Resources for successful conservation management

Various resource implications (in the broadest sense) follow from this for the successful management of designated conservation areas. Firstly, ecological research needs to be brought into closer connection with the practical needs of conservation in specific areas of the globe, and, as many ecologists and con-servation managers are currently arguing, this means a change in the institutional arrangements within the academic world, in which such applied research is looked at askance (Rogers 1997: 60–77). This also suggests the need for conservation managers themselves to be continually updated on the latest developments in ecology. Rogers (ibid.: 70) further suggests the desirability of a 'pragmatic tech-nological interface' between science and conservation practice, perhaps modelled on the role of commercial pharmaceutical companies acting as intermediaries between medical researchers and medical practitioners. This is an idea which fits readily into the doctrine of 'ecological modernisation', of course. It envisages a new market niche for commercial enterprises to turn the findings of pure eco-logical research into usable products for the attainment of practical conservation ends. In the case of conservation, this could perhaps be provision of intangibles, such as processes for the acquisition of useful ecological information, techniques for conducting surveys and monitoring crucial environmental changes and so on, rather than pieces of hardware (ibid.: 71).

Secondly, the conservation of species and habitats within a specific context is a practical task. Managers need to have achievable and effective courses of action open to them, hence the need to ensure that they posses the decision-making skills needed to order priorities within complex situations of uncertainty (Possingham 1997: 298–304). Given the importance of knowledge of the specific context represented by a specific conservation area, it is arguable that managers should be given the discretion to exercise their own judgement about how best to pursue their conservation goals, rather than be required to conform to some detailed central directive from the body with overall administrative authority for the conservation efforts within state boundaries. As Kelsey has noted, the preservation of biodiversity requires the concept of 'adaptive management', which, he tells us:

> was first defined by an interdisciplinary team working at the International Institute of Applied Systems Analysis in Laxenburg, Austria, in the mid-1970s. Instead of assuming that all eventualities can be forecasted and planned for, adaptive management is based on the belief that surprises are inevitable, and thus policies and organisational structures should be flexible and responsive to change.
>
> (Kelsey 2003: 383)

However, from the point of view of ecological justice it will be crucial that the preservation of endangered species, where these are at issue within a given conservation area, be put automatically at the top of the goals to be pursued within that area (as opposed to recreational, economic, cultural or other goals legitimately pursued within conservation areas).

This brings us to another desideratum for the institutional arrangements governing conservation management, namely that measures be taken to train managers in the necessary skills for interacting with those who inhabit the conservation area, the general public, the media and other decision-makers and stakeholders. As Meyer puts the point: 'Conservation is essentially management of human activity in the landscape, so to ignore the societal context for conservation efforts is to invite failure' (Meyer 1997: 141). Conservation is, thus, an inherently political activity, and its practitioners need to be trained in the skills of interaction with those human groups whose demands and criticisms will provide the inevitable context of management activity. Arguably they also ought to be furnished with systems of support and advice to allow them to perform this task well.

The role of other stakeholders

One of the implications of the new paradigm in ecology is that it is impossible to cordon off conservation areas hermetically. The activities of groups of people adjacent to conservation areas will become of crucial importance. In other cases it may be possible to protect only a part of the area crucial to the survival of a species of organism, and other areas may lie outside the zone of protection, perhaps parts vital to the replenishment of that within the protected zone if it should suffer some unforeseen depredation, as in the case of those patchy distributions of some species which have come to be known as meta-populations (see Wiens 1997: 99). Sometimes private land will lie within a protected area, and the inhabitants will be permitted to go about their normal activities. In cases such as this, as has sometimes been noted (Pulliam 1997: 17), it will be crucial to the success of the conservation effort that those who own, individually or jointly, private land be rewarded for playing their part in protecting habitats and endangered species.

These points apply both in developed countries, where the concept of the private ownership of land is fully developed and fiercely defended, and in developing countries, where the landholding stakeholders are in possession of communal lands and are less able to mount robust legal defences of their rights. In addition, account has to be taken of the increasingly widely held and well-supported view, promulgated by such groups as the Forest Peoples Project, that the forcible removal of the long-standing local population from the areas earmarked for conservation purposes has been both inhumane and, probably, unjust for those people and ecologically damaging for the species which this manoeuvre was designed to protect (Forest Peoples Project 2003). There is some hope in the fact that some recent studies have shown that, where indigenous peoples are left in possession of their traditional lands, scientific and traditional knowledge bases can be integrated to the benefit of all interested parties, provided that a fully participatory and transparent form of joint decision-making in the management of the lands is developed (Kelsey 2003: 392). There is nowadays a clear recognition in most conservation circles, including the parties to the 1992 Biodiversity Convention, that the knowledge base of indigenous peoples living in areas of conservation importance should be respected, both as a part of the necessary respect which the peoples themselves are owed, and as an important conservation resource (Potvin *et al.* 2002: 159).

However, it has also to be recognized that the beliefs, values and traditions of such peoples do not necessarily coincide with the aims of protection of local populations of non-human species, even endangered ones. For example, the hunter-gatherer Embera people of Panama give names only to species of plant and animal which are of practical use to them (Potvin *et al.* 2002: 167); they will without hesitation kill specimens of what scientists have classified as endangered species which they see as threats to their interests; they explain rarity of animals not as a result of their own hunting practices, but in terms of their own cosmology, according to which increased rarity may be as a result of the animals in question taking refuge in the Netherworld, conceived of as 'a place of refuge and rest for the animals' (ibid.: 170–1). What such examples should lead to is not, of course, the setting aside of indigenous peoples' beliefs and values as worthless, but a recognition that the bringing of such beliefs and values into fruitful contact with the perspectives of conservation practices is going to be, often, a fraught and delicate affair, requiring conservation managers to show forbearance, sympathetic understanding of indigenous peoples' world-views, and creativity in reconciling the needs of conservation and the practices of such people (ibid.: 172).

From the point of view of ecological justice, of course, it will be a requirement of justice that owners and users of land exercise the necessary care and forbearance to ensure that viable populations of other species are preserved. However, it is a basic requirement of justice too that, where a moral responsibility which basically applies to the whole of society falls specifically upon one group of people, they be given help and support to bear it. It will be both inequitable and ineffective simply to remind them of their moral duty and then leave them to get on with it. It will, of course, be incumbent upon state authorities to ensure that such people are given information and support to carry out their management of their resources in such a way as to satisfy the requirements of ecological justice. The reward in question may have to take the form of monetary compensation if the individuals concerned are prevented from managing their property in a way which serves their interests but puts the continued existence of the species in question under threat.

There are many precedents for interfering with the individual right to property in this way so as to secure morally defensible ends which are important and not attainable in any other way. An example from the UK is the legal requirement that householders who discover a bat inhabiting their roof space are required to leave it undisturbed, unless it invades their habitation area, that the roost area itself is protected by law, even if no bats are in residence at a given time, and that permission must be sought from the appropriate official conservation body if one wishes to do something to one's house, such as alteration or maintenance work, which might affect bats or their roosts (Scottish Natural Heritage 1999: 18–19). This may not be an easily enforceable law, but it does suggest that human beings in at least some jurisdictions, in pursuit of the protection of the interests of non-human species, have already rejected the claim that people's property rights are inviolable.

The more general point which emerges from these examples, however, is that successful protection of endangered species requires that people who are not conservation professionals, and who control habitats and other resources necessary for the continued existence of such species, must be informed about the ecological situation, fully consulted in the decisions about how to address the conservation issue, persuaded to participate fully in the protection measures, compensated for any sacrifices on their part required by this effort, and rewarded, if only with public praise and recognition, for engaging in it. These efforts fall upon government bodies, which must accordingly be active in ensuring that such individuals are not ignored or ridden over roughshod in the pursuit of ecological justice. But this assiduity in bringing all relevant stakeholders on board must be matched by a firm resolve that ecological justice be done, in spite of the conflict of interests that may emerge in the course of its attainment.

The role of the media and education

It is a frequent complaint of conservation managers, ecologists and environment-ally concerned groups of all kinds that the mass media do not give much prominence to environmental issues in general unless there is a dramatic crisis story to relate, that professional journalists have very little incentive to become experts or specialists in the field of the environment, and that issues which are reported are often misreported, intentionally or otherwise. In addition, groups which are hostile to environmental protection are often adept in their media activities at misusing scientific findings and at presenting unrepresentative views from within the scientific community as if they were widely accepted (Pulliam 1997: 21; O'Neill and Attiwill 1997: 351).

Given the nature of the media market-place, in which news stories are basically a device to attract readers to a place where they can be exposed to commercial advertising, it is more than usually the case that a set of requirements for the improvement of this state of affairs has the character of a wish-list. Thus, it is not difficult to see that commercial press and broadcast media ought to have on their staffs journalists with the knowledge base to write competently about environmental issues and that the media themselves ought to give the issues the prominence which their importance merits, and so on. From the point of view of ecological justice, it will also be very important for journalists to have an understanding of the moral dimension of the issue of species protection, and the capacity to present it accurately, even if they are personally sceptical about the cogency of the moral arguments in this area. However, it is not clear that any of these requirements can be imposed upon a press in a liberal society, and so achieving their implementation can only be a matter of concerned publics pressing for such developments, perhaps via the exercise of their consumer sovereignty in the market-place.

In the case of public service broadcasting it may be easier to argue that the broadcasters have a duty to reflect the moral consensus of their society, as well, of course, as providing a forum for critics of that consensus to mount challenges. But it is clear that even in this area such a development will have to be consequent upon the prior emergence of such a consensus. Defenders of ecological justice will first have to make the case and gain widespread acceptance for it before they can press a 'consensus' case against public service broadcasters. This does mean, of course, making full use of the opportunities for critique of the status quo which, as has just been noted, ought to available as a matter of course in the provisions of public service broadcasting in a liberal society.

From the other side of the issue, it will also be desirable for those responsible for the institutional implementation of ecological justice to be able to present issues effectively to the media, and thereby, of course, to the general public. The latter in turn ought to receive in the course of their education a schooling in environmental issues in general and a grasp of the moral component of human–nature relationships, with an explanation of the arguments in favour of ecological justice. Recently there has developed an interest in the possibility of environmental citizenship education (Dobson 2004). States which do embody constitutional and institutional protection for non-human species will certainly need to ensure that their citizens remain fully aware of the reasons for this. The case for extending justice to non-human creatures will thus form part of the standard content of education in a state which embodies the sort of ecologically just liberal regime argued for in Chapter 8.

The role of governments

Governments contain the prime decision-makers in all environmental areas. The case has already been made for states to embody an ecologically just form of liberalism. What can actual experience of conservation management tell us about the institutional problems we can expect in a government of such a state? The key requirement is that legislation to secure ecological justice, and in particular to protect endangered species, should be informed by the best available scientific understanding of how such protection may be effected, in the light of the new paradigm of ecology outlined above. The key to this is that protection must be understood as an activity which has to be integrated into a wide variety of areas overseen by governments, and not simply pigeonholed in a box marked 'conservation areas'.

Governments have to legislate, establish bureaucracies to enforce the legislation, provide monitoring and research activities and oversee the work of conservation managers. The problems already outlined indicate the constraints under which such institutional arrangements should operate to secure efficient protective arrangements. There needs to be inbuilt flexibility in the work of conservation managers, who need to receive the training, resources and support mentioned above. Governments can encourage and support the 'pragmatic interface' organizations whose usefulness to conservation managers has already been noted. The oversight department responsible for conservation needs to keep a critical eye on the impact of social and economic developments more generally within the state and beyond its borders upon the conservation goals set out by the legislation.

A state which has committed itself, preferably by constitutional amendment, to the goal of ecological justice will need to work out how to secure, within conservation practice, the goal of preserving the species which exist within its jurisdiction, as well as how to play an effective part in pursuing this goal in international forums. In this regard, within the conservation practice for which it has direct responsibility, it will give the highest priority to the location and protection of endangered species which are endemic to its territory. It may be that the territory of a given state will contain no such endangered endemics. The next priority in that case will be to locate any endemic species not currently endangered, but under threat if certain developments occur, and strive to ensure that those threats do not materialize.

Some states may contain no endemics at all within their borders. That is, even if all their species were wiped out – locally exterminated – those species would still survive elsewhere. However, it will be recollected from the arguments put forward in Chapter 9 that governments still have the responsibility to ensure that the claims in justice of viable local populations of organisms to their fair share of environmental resources are taken account of in human decision-making. The case for overriding such claims has to be made out in accordance with the criteria of ecological justice presented earlier. Where the species in question become individualized to a higher degree, these claims increasingly reside in the individual members of the species. Thus there is a prima facie case in ecological justice for local populations, and individuals, of even non-endemic, non-endangered species to retain access to the environmental resources they need to survive and thrive. In this way, arguments based on ecological justice go beyond the goal of maintaining biodiversity, which may be satisfied, perhaps, if thriving populations of all species exist somewhere on the planet, albeit in a highly localized manner.

In the case of clashes of interest between human beings and such populations and individual members of species, the human need has to be sufficiently weighty to overturn the claims of the non-human species. In the case of non-endangered non-endemics of highly non-individualized species this will often not be an onerous requirement. But, given the new paradigm of ecology, decision-makers will have to be sensitive to the possible further implications of any modifications to the environmental patches which they are proposing. That is, even the local eradication of the population of a species which is not endangered may have serious ramifications for the wider environment, perhaps pushing other species, including endemics, towards extinction.

All of this looks to be placing an onerous requirement upon the governments of states. Their conservation-oriented activities must be wide-ranging, ever-vigilant, and continually questioning of the implications of the environment-altering activities in which their citizens engage. It has to be said, however, that this onerousness is present to a high degree even if the case for ecological justice is rejected. The new paradigm of ecology displays as never before that serious efforts to preserve other species, even for the most anthropocentric of reasons, such as the possible economic, scientific, cultural, recreational, spiritual and aesthetic benefits they afford human beings now and in the future, will require governments to make much more strenuous efforts than any is currently making.

In a liberal society, with a lively body of NGOs committed to making governments attend to such issues, there is still reason to hope that these requirements can be developed in a rapid enough manner to do some good. The power of moral argument ought to help here, and the theory of ecological justice is intended to develop the strongest form of such moral argument.

Having now outlined some of the crucial ways in which institutions within an ecologically just liberal state might derive from existing, albeit imperfect, conservation institutions, we need to consider the inter-state, or international, dimension, which, as noted at the start of this chapter, is going to be crucial for the successful prosecution of the aims of ecological justice. As in this chapter, we will in the next seek to develop and extrapolate from a present complex and unsatisfactory, but also in some respects promising, reality, rather than attempt to outline a utopian ideal.

12 Institutional arrangements at the global level

As many analysts of international environmental institutions emphasize (for example, Haas *et al.* 1993: 4; Hurrell and Kingsbury 1992: 6–8), the political structure of the world for the foreseeable future will be conditioned by the continued existence of states and their commitment, albeit nowadays somewhat nuanced and circumscribed, to the ideal of state sovereignty. This means that proponents of ecological justice will need to pursue their aims within a system of developing world governance involving cooperation between state-based executives, rather than government involving a single global executive. The development of international environmental institutions (IEIs) oriented towards ecological justice concerns can, therefore, come about only as the result of actions undertaken by, and agreements made between, state governments.

The traditions of diplomacy and the perennial attractions of realist ideologies of state actions at the international level may seem to make such a development a remote prospect. However, it is important to see (as Haas *et al.* 1993 reveal) that, once IEIs and other international governance institutions have come into existence, then this itself makes a significant difference to the conduct of international politics. For the personnel of such institutions can themselves play a key facilitating and agenda-setting role in their own further development and in the creation of new forms of international institution.

For them to do so, as Haas *et al.* argue, various factors need to be in place. It is important for the IEIs to be small enough for their own administration not to be a time-consuming distraction. They need to be alive to the need to create networks of interest with other IEIs, other international institutions, state government actors and bureaucrats, and personnel of NGOs within states. It is very helpful if the terms of reference of IEIs are drawn sufficiently wide to allow growth of activity and regulative competence in previously unforeseen areas. Even the fact that an IEI may initially be established on the basis of only a vague commitment to some general set of principles can be a strength rather than a weakness, as it may allow the acceptance of IEIs by states which would otherwise be hostile, and give the IEI in question breathing space to move towards a more precise and useful set of regulatory actions. However, it should be noted that Haas *et al.* discovered that IEIs tend not to enforce rules, but rather put pressure on states to alter their behaviour by monitoring their performance in a given area of environmental concern and liaising with domestic NGOs to help them press the case for more effective action by their government (Haas *et al.* 1993: 408–15).

Given such possibilities, and in spite of the continued and entrenched position of the 'sovereign' state as the key international actor, when we examine the actual history of the development of international institutions since the Second World War we find many reasons for hope from the point of view of ecological justice. Firstly, in spite of the obvious problems of achieving agreement between different nations and cultures over matters of value and ethics, the world has in fact witnessed during the last half-century the development of a plethora of institutional arrangements for international governance. This development encompasses arrangements for international debate and decision-making on matters of widespread, and sometimes of universal, concern. The UN Security Council and its resolution-making apparatus is the most obvious example to mention.

These developments, of course, may still be subject to purely realist interpretations. Thus the ethical debates in governance institutions which centre on such notions as justice and human rights may be regarded as essentially rhetorical devices for disguising naked self-interest. Nevertheless, it is enormously important, both that such debates take place, and that it is an ethical window-dressing that is sought (if that is what is going on), not least from the standpoint of this book's moral universalism. Participants in such activities are at least paying lip-service to the need to work towards a universally accepted position of fundamental moral matters.

The idea that all politics is about self-interest is an ancient one, and one which perennially recurs whatever the level of politics we consider. Its plausibility, however, is only superficial. Human action, at the individual level, and at the group level where problems of collective action raise their knotty heads, may be largely self-interested. But ethical motivation is present too, even if only fitfully, and is at least perceived by most people as the only finally justifiable form of motivation. Hence the sheer existence of international governance institutions with a clear, if problematic, moral dimension is the first reason for hope from the point of view of proponents of ecological justice.

Secondly, and more pertinent to the matter of ecological justice, is the fact that of the two currently operative focuses of international relations – international security and sustainable development – the latter is already encompassing a whole swathe of international framework conventions and protocols which focus upon global environmental issues, including the one which is most closely related to the matter of ecological justice – the Biodiversity Convention. It is also pertinent to note that these conventions are not regarded as in any way an insignificant part of the international system of governance. As McGraw has noted, 'Together, the three "Rio Treaties" enjoy a special status within the UN system, since they are among 25 treaties identified in the Secretary-General's Millennium *Report* as central to the UN's mission' (McGraw 2002: 23).

As a recent example of promising developments in this area one can cite the sixth meeting of the Conference of Parties to the UN Convention on Biological Diversity which took place at The Hague in April 2002, at which a strategic plan was produced containing the aim of achieving significant reduction of the current rate of biodiversity loss at global, regional and national levels by 2010. This aim was endorsed by the Hague Ministerial Declaration on 14 April 2002 and by the World Summit on Sustainable Development held in Johannesburg in September of that year. The UNEP-initiated Millennium Ecosystem Assessment, which began in 2001 and continues for four years, is due to provide 'the most extensive study ever of the linkages between the world's ecosystems and human well-being' (Millennium Ecosystem Assessment 2003), and will provide crucial data for the Biodiversity Convention and for three other conventions important for biodiversity, namely those on wetlands, desertification and migratory species.

Granted, the whole sustainable development discourse which currently predominates is completely anthropocentric in its ethical approach, and biodiversity protection is thus defended on the basis of purely human benefits – economic, recreational or spiritual. Nevertheless, there is now a significant element of international effort directed upon the relevant matter for concern, from the point of view of ecological justice, even if not for completely the right reason. If all goes according to plan, then by 2010 the world, under the auspices of the UN institutions and conventions, should have an excellent idea of what precisely is happening to the planet's ecosystems and the species which they contain, and have taken steps to reduce 'significantly' the current rate of loss of biodiversity.

Another recent development which adds a desirable dimension to the international attempt to preserve other species was the creation by UNEP and UNESCO of the Great Apes Survival Project Partnership (GRASP) in 2001. As the name suggests, the aim of GRASP is:

> to lift the threat of imminent extinction faced by bonobos (pygmy chimpanzees) and gorillas, and serious threats to chimpanzees and orangutans. Beyond that our mission is to conserve viable, wild populations of every kind of great ape, and to make sure that their inter-actions with humans are mutually positive and sustainable. We also seek to exemplify and relieve the threats faced by other kinds of animals, birds and plants sharing the forests where apes survive, and to illustrate what can be achieved through a genuine partnership between all the stakeholders in these fragile ecosystems.
>
> (GRASP 2003: 1)

As we saw in Chapter 10, the case of the great apes is one in which the idea that 'human rights' may properly be promoted beyond the purely human sphere is at its most immediately persuasive. The UNEP initiative does not reach that point, however. There is acknowledgement that:

> Great apes are highly intelligent, social primates who share many highly valued qualities with humans; they can reason and communicate emotions such as joy and grief; those in captivity have mastered some form of language . . . and apes make and use tools in the wild. Great apes and humans are the only species to be able to recognise themselves in a mirror, a sign of a highly

developed self awareness. Great apes also have the capacity for complex social interactions.

(GRASP 2003: 1)

However, none of these points is used to argue to the conclusion that the great apes have a right in justice to a share of the environment so as to live in accordance with their own natures and purposes.

The close evolutionary connection with human beings is also noted, and the conclusion is drawn that 'great apes form a unique bridge linking humans to our ancestors', thus: 'If we were to lose any of the great ape species many people would feel that we were destroying a part of the bridge to our own origins, and with it a part of our own humanity' (GRASP 2003: 1). Here too, therefore, the human connection is used, not to accord the great apes their own moral standing, and thus to argue for the existence of human moral obligations towards them, but rather to give yet one more reason why it is in the interest of human beings that they should be preserved.

From the point of view of ecological justice, therefore, GRASP embodies a defective approach. But it is significant as edging towards the kinds of consideration employed in the creation of the case for ecological justice. As such, it makes a significant advance, for it is an initiative with the organizational and financial backing of the world's most important institution of international governance which seeks to preserve viable populations of a non-human species, and which does so by appealing to features of that species that can be used to ground a persuasive case for what is recognizably an example of ecological justice.

We will consider at a later point the implications of this initiative for the development of international institutions of ecological justice. But first we need to examine the prior issue of how such institutions may be brought about.

How are international environmental institutions created?

To do this it will be useful to introduce into the discussion some basic but key distinctions, as provided by Haas *et al.* in their well-known study of the development of IEIs (Haas *et al.* 1993). Institutions may, of course, be full-blooded organizations, but the term can usefully apply also to conventions and regimes, in which patterns of behaviour become standard without the underpinning of a bureaucratic organization designed specifically to encourage or enforce them (ibid.: 5).

In their analysis of IEIs, Haas *et al.* work with the standard division of environmental issues between transboundary, global commons and intra-state issues (Haas *et al.* 1993: 9). The first of these concerns adverse effects produced beyond the boundaries of a given state upon the environments of one or more other states. The second concerns adverse effects of state actions (and actions of non-state actors protected by some state or other, such as the activities of deep-sea fishermen) upon the environment beyond state jurisdictions. A better name for these issues would be 'open access', since, as many have argued, commons are historically areas where access has been regulated by some organization, state-sponsored or otherwise. The third concerns issues which affect only the environment within a given state's boundaries.[1]

This tripartite distinction is important for Haas *et al*.'s analysis, for they find that IEIs are best placed to affect the first two kinds of issue directly (Haas *et al*. 1993: 14). IEIs may still have a part to play in providing monitoring services and support for groups within a state which are seeking to get it to amend its environmentally harmful activities. But the key role in such issue areas will always be played by the internal institutions of states. The analysis given in the last chapter of the domestic structure of institutions needed to put into effect ecological justice thus deals with the paramount institutions for this issue area. However, the adverse effects upon non-human species which are the concern of ecological justice may also be found in the other two issue categories – transboundary and open-access. States may act unjustly towards species in other states and in open-access areas.

It is in these areas that IEIs appear to be most obviously required. It is here, too, that the perennial problems of collective action raise their messy heads. The key move in developing workable IEIs will thus usually turn on giving all participant states confidence that their efforts, often involving sizeable expenditure of time, money and effort, will not be taken advantage of by actors, state and otherwise, who behave as free-riders. Proponents of ecological justice, therefore, in their attempts to propound remedies for transboundary and open-access injustices to other species, will need to be aware of the need to deal with such assurance problems. There is evidence that such outcomes are already taking place, with the creation in southern Africa of 'peace parks' which link together national parks in adjacent states in order to create a super-park, giving the species within them much greater freedom of movement across state boundaries than hitherto (Peace Parks Foundation 2003).

Interestingly, however, although Haas *et al*. found that IEIs can more easily directly affect issues in the transboundary and open-access areas, they also found that success even in these areas is consequent upon an effective group of actors – 'concerned publics', NGOs, bureaucrats, the media – operating within the boundaries of states. As they claim:

> *If there is one key variable accounting for policy change it is the degree of domestic environmentalist pressure in major industrialized democracies, not the decision-making rules of the relevant international institution.* However, we do find that the institutions that have given rise to the most dramatic changes in collective policy-making are those that were able to apply constructive channels for such domestic pressure to reach governments.
>
> (Haas *et al*. 1993: 14; original emphasis)

This suggests that, even in the areas which appear prima facie to be entirely international in character, the crucial issue will concern the domestic politics of the appropriate states. Once again, this takes us back to the internal political order of states and to the need to foster the kind of domestic political arrangements outlined in the preceding chapter. The crucial issue from the point of view of this study can be highlighted when we introduce a further tripartite distinction from Haas *et al.* This concerns the three areas of policy-making and implementation in which, in their view, IEIs can engage. These are what Haas *et al.* label the 'three Cs', namely concern, capacity and contractual environment (Haas *et al.* 1993: 11). They cover the concern which a government has to deal with an environmental problem, its administrative and other forms of capacity actually to do something effective, and the general climate, internally and externally, for a government to make and keep international agreements – with the possibility of dealing with the aforementioned collective action and free-rider problems being paramount.

Of these three elements, the first, governmental concern, is vital (Haas *et al.* 1993: 19). Unless a government can be induced to be concerned about an environmental problem it will not be inclined to push for the creation of an IEI to deal with it and will be a reluctant or even obstructive force when other states attempt to address it on the matter; even if induced to sign up to an IEI, it will be, at best, a candidate for laggard status or, at worst, be a free-rider. Without such concern on the part of governments, the ability of IEIs to help a government to increase its capacity to implement the requirements of an international regime and its ability to overcome collective action problems by fostering a beneficial contractual environment will be of little use.

The implication of these claims is very significant. They suggest that it is a necessary, though not sufficient, condition for success in the creation of effective IEIs that states whose actions are crucial to the success of IEIs are such as to permit their domestic publics to mount effective pressurizing campaigns on their governments. This in turn once again underlines the importance, especially from the point of view of defenders of ecological justice, of supporting liberal forms of democracy within the world's states. For only such democracies, with active and effective civil societies, contain the permanent possibility of concerned environmental activists, including the proponents of ecological justice, to act so as to generate the appropriate levels of concern – without which, as we have seen, IEIs of whatever kind are unlikely to succeed. Hence we have a further, powerful, reason for supporting the connection between ecological justice and the form of liberalism developed earlier in connection with Barry's conception of justice as impartiality.

It is true that, as a matter of happenstance, a wholly non-democratic state – authoritarian or totalitarian – may be governed by people who are supportive of environmental purposes in general and even of ecological justice in particular. But the standard liberal arguments against 'benevolent dictatorships' apply as strongly as ever. The case also has to be made by defenders of ecological justice against the pre-liberal idea of 'republican democracy' which has recently attracted the favourable attention of conservative communitarians (see, for example, Cantori 2003). They argue that, in a state whose citizens are united by a common allegiance to a time-honoured religious world-view, such as Islam, the need for, and possibility of, a vibrant civil society such as characterizes the atomized and individualistic societies of the secular West does not exist. Instead, such states are required by their citizens to embody and protect a view of the 'good life' which requires various forms of virtue among the citizenry and state personnel, and a concomitant need for the appropriate forms of political space within which the informed citizenry can be critical of the government and hold it accountable. Clearly, there is the abstract possibility that such a republican democracy may be committed to a world-view which is hospitable to the requirements of ecological justice. However, this does not seem particularly probable of the world-views which we are likely to be faced with in practice in such republican democracies, hence a proponent of ecological justice should not be seduced by this non-liberal alternative.

In any case, there are strong liberal arguments against republican democracy, and the communitarianism on which it rests, such as that, in the absence of a vibrant civil society, the critical role of the informed, virtuous citizen is highly problematic and that republican democracy really only provides us with a recipe for simple oppression by the guardians of the state-sponsored 'good life'. Also, as we saw in the course of considering the arguments of Barry in an earlier chapter, arguably no state is justified in embedding a view of the 'good life' within its constitutional arrangements, even if it has the complete support of all its citizens at a given time to do so, for this is to act against the possibility of individual dissent from the shared view and thus is *ipso facto* an indefensible attack upon the rights of individuals to reject majority views as they see fit. Of course, defenders of republican democracy may deny that such rights should be recognized, but the dissenters have a conclusive reason for thinking otherwise, namely that it is their autonomy which is under attack, and this point leads then directly to the arguments which Barry elucidates with use of the concept of the 'agreement motive' discussed earlier. Further, no world-view, however ancient and hallowed, is devoid of interpretative possibilities which will inevitably lead to dissent even between people who regard themselves as supporters of that view. Hence a republican democracy is always inherently unstable as well as offensive to the most defensible conceptions of human rights.

The upshot of these reflections is then that, from the point of view of a defender of ecological justice, a world of states embodying the basic elements of liberal democracy, especially civil society institutions such as NGOs, is essential to the realistic pursuit of such IEIs as are needed for the implementation of such justice, whether we are considering purely state-bounded, transboundary or open-access forms of ecological injustice. The problematic aspect of all this is, of course, that liberal democracies are battlegrounds of rival value and other positions. The existence of a civil society may give defenders of ecological justice the possibility of pushing their state government towards a concern with ecological justice, and to embed moral responsibilities towards non-human life-forms within the state's constitution. But plainly that is a difficult course of action given that other groups may well be actively hostile to the whole idea. Still, as we saw in our discussion in Chapter 7 of Bell's version of political liberalism, this route, difficult and disappointing as it inevitably is, does seem to be the only one open to the defenders of ecological justice, as indeed it does to proponents of all the other forms of distributive justice with which we are familiar.

We are now in a position to say a bit more about how those IEIs that currently exist typically came into existence, and thus to speculate in a more realistic way about how IEIs needed for the purposes of ecological justice may come into existence, at least in principle. However, it has to be emphasized that what follows is, in the nature of the enterprise, highly speculative, even if informed by the actual history of international political institutions.

The history of such institutions, such as the UN, since the middle of the twentieth century, fraught though it is, suggests that it is possible for there to emerge a consensus between at least a significant number of key states that they are faced with an important and pressing need which they can satisfactorily address only by undertaking joint action. The need for security is the most time-honoured of these, and is the most readily intelligible, given the universal and long-standing propensity of human beings to resort to deadly force to resolve disagreements and clashes of interest between themselves. The emergence of the sustainable development paradigm is a much more interesting case, resting as it does on the acceptance of some recondite and often controversial scientific theorizing, much of which involves the tracing of unobvious (at least to the layperson's eye) inter-connections between large-scale and hard-to-observe phenomena.

Of course, many environmental critics would maintain that the concept of sustainability has been purloined by international business groups to subserve the interests of capital. Even if this is so, it is nevertheless of considerable significance that the claims of those who first raised concerns about the sustainability of human economic activity have had to be met by those who defend the dominant capitalist market system, rather than simply be ignored or ridiculed (although many powerful groups still exist which try both of these tactics, albeit with declining effectiveness). This suggests that human beings and the states they represent at least have the capacity to consider and accept claims and ideas which go beyond the blindingly obvious, and to undertake joint activity in pursuit of some difficult and controversial aims.

However, in the case of ecological justice, the difficulties faced by the propo-nents of sustainable development are compounded by the fact that the key concept moves resolutely beyond a purely anthropocentric moral position. To gain acceptance of this among those not already strongly inclined to accept it requires the propounding of some fairly recondite moral theorizing. In a world in which many hard-headed people, especially at the international level, believe themselves to have quite enough to do to advance the two paradigms already mentioned, the willingness to contemplate such theorizing may be hard to find.

How may such a willingness be generated? In terms of conventional analysis of the development of international regimes and treaties, we need to look for various desiderata in order to gain the successful implementation of a regime of ecological justice (Hurrell and Kingsbury 1992: ch. 1) The matter has to be put onto the international agenda, which requires an agenda-setter. The agenda has to be pursued by some influential actors on the international stage – there have to be 'lead' states or actors of some kind. As we have already noted, the upshot of Haas *et al.*'s analysis of a variety of existing IEIs is that the existence of 'concerned publics' within liberal democracies is a necessary condition of the development of sufficient concern among possible lead states to seek to put the matter on the international agenda. To use the terminology provided by Porter and Brown (1996: 32), undoubtedly such lead actors will be faced by opposition states and groups who will seek to exercise a veto on the development of this new paradigm – a veto coalition. A decision will often turn on the attitudes and votes of previously uncommitted groups and states – the swing group. To put the point succinctly, international regimes often emerge as the result of lead states offering inducements to the swing group to support them in the overthrow of the blocking attempts of the veto coalition.

Let us, then, consider the possibilities for agenda-setting, lead states, veto coalitions and swing groups. Agenda-setting may emerge first as the result of non-governmental group activity within certain states. One recent example is that mentioned in the previous chapter, namely the emergence in Germany of a constitutional commitment to the protection of animals. More generally, the animal rights movement has had some influence and success in many states across the world, primarily in the development of legislation to protect animal welfare in specific regards and contexts, rather than in the form of entrenched constitutional protection on the German model (Singer 2003). Perhaps most germane of all to our concerns is the important initiating role undertaken by the World Conservation Union (IUCN) in the development of the UN Biodiversity Convention. This shows that it is possible in certain circumstances for an NGO with sufficient prestige to have a direct impact upon an IEI, for the draft biodiversity convention which the IUCN drew up in the early 1980s attracted the interest of the United Nations Environment Programme (UNEP), as well as of various states (McGraw 2002: 10).

These forms of legislation, and the actions of animal rights advocates based on ethical and philosophical analyses, all represent crucial movements away from the purely anthropocentric moral standpoint. They embody the idea that at least some non-human beings possess moral standing in their own right. There are some ways in which they may not be treated, however advantageous that treatment might be to human beings. It is thus not wholly far-fetched to believe that the case in justice for non-humans to have a claim for their fair share of the earth's environment may gain acceptance within such prestigious environmental NGOs as the World Conservation Union and receive a hospitable welcome within possible lead states such as Germany, some Scandinavian countries and New Zealand to the degree that the already existing intra-state institutions mentioned in the last chapter may be developed so as to encompass the aim of ecological justice.

The next step, of course, is to envisage such states attempting to put onto the international agenda such moral concerns and arguments. According to some analysts of international politics, such as Callicott (2002: 18–20), the idea that nature possesses intrinsic and not simply instrumental value has already gained acceptance in the corridors of international decision-making. In support of this claim, one might cite, for example, the preamble to the 1992 UN Biodiversity Convention, which opens with the phrase: '*Conscious of the intrinsic value of biological diversity* and of the ecological, genetic, social, economic, scientific, educational, cultural, recreational and aesthetic values of biological diversity and its components' (Le Prestre 2002b: 345; my emphasis). Thus, while it is clear that the aim of preserving biodiversity in some sense has already been granted acceptance within the institutions of international governance mainly in terms of human benefits from the exploitation of gene pools and so forth, it does seem that the idea has also been accepted that the existence of other species possesses a value which goes beyond any conceivable human uses.

Moving on to the question of which states are likely to form blocking coalitions and swing groups, the 1992 Biodiversity Convention indicates one obvious set of possibilities. The concern to protect biodiversity there emerged as a common concern, but one which was differently understood by different states. The life-forms which were the main focus of concern are to be found largely in the states of the global South. This is certainly true of the great apes, which are located in certain states of sub-Saharan Africa and South-East Asia. But the main thrust of the concern came from the global North, where a lot of the interest in using organisms as resources, particularly in the development of agricultural and pharmaceutical products, was very strong. Concerns about the health of ecosystems and the growth of scientific knowledge also played a part in the arguments for the convention. But a crucial practical problem focused on the conditions in which the poverty-stricken South would be prepared to preserve habitat and allow the northern companies access to the biological resources within their territories (McGraw 2002: 7). These are issues of distributive justice within the human species. One may expect them to be resolved on the basis in part of moral principle, and in part of bargaining, as happened in the case of the Biodiversity Convention. This, under the pressure of the South's development concerns, emphasized that biodiversity should be seen as a resource and that sustainable use and beneficial sharing should be set forth as aims of the convention in addition to conservation, about which the North was most concerned (ibid.: 32).

The case for ecological justice will also face a similar issue. The bulk of species, and the bulk of endangered endemics, on the planet are to be found in the global South, especially in those 'hotspot' areas of the tropics in South America, Africa and Asia. The preservation of species will thus require expenditure of time, money and effort within those states, and the possible forgoing of certain kinds of economic activity. There is no obvious reason why the governments or peoples of such states will be particularly resistant to the moral case put forward by ecological justice proponents, indeed some of the local cultural formations within those countries might make them more open to the moral case than is the situation among the inhabitants of the global North. But the question of who pays to secure that justice, which might on all hands be agreed to be an important desideratum, will obviously remain the key issue even if biodiversity protection is viewed in moral and not just prudential terms.

The obvious and most defensible answer is that those who can afford the most to secure such justice must be prepared to pay the most. This sets up a differential set of responsibilities – the rich North will have responsibilities not held by the poor South. Arguably, however, this principle has already been accepted since the onset of the sustainable development agenda. As Paul Harris has argued (2001: 244–54) northern states, especially the USA under President Clinton, recognized that the North should pay more, and have different responsibilities with respect to environmental protection, than the South. A principle of international equity, in other words, has already been accepted, although it cannot yet be said to be a permanent element of international decision-making. However, the issue of who pays will undoubtedly be the basis of at least some blocking moves within the South, until the equity principle is accepted here as elsewhere. This is, in other words, the area where ecological and human distributive justice intersect.

It has to be said, however, that in spite of these difficulties there currently exist over ninety IEIs which between them cover a great many species and seek to protect them against the depredations of human beings. The five major IEIs in this area are the Convention on International Trade in Endangered Species of Wild Fauna and Flora (CITES); the Convention on the Conservation of Migratory Species of Wild Animals (CMS); the Convention on Wetlands of International Importance, especially as Waterfowl Habitat (Ramsar Convention); the Convention concerning the Protection of the World Cultural and Natural Heritage (Paris Convention); and the UN Convention on Biological Diversity (CBD). Taken together these have been identified as a crucial developing source of international law aimed at protecting biodiversity (McGraw 2002: 10). Hence, we can say that there is already a great deal of protection of non-human species going on – in addition to the measures enacted by the governments of particular states to protect species wholly within their borders.

The existence of this extensive set of IEIs and their domestic equivalents suggests at least one possibility for the development of ecological justice IEIs, namely that their work could be given a new rationale: the pursuit of ecological justice. In some cases – the CITES convention, say – this might not be a difficult transition to effect. In the case of IEIs designed to protect species for their economic use for human beings – such as fisheries protection – it may be more problematic. But even here, given that ecological justice is compatible with at least some use of other species as resources, involving their killing, this may not be too much of a leap to make.

Of course, what may emerge as the result of all the real-world political activity we are contemplating here may be something worthy and high-flown but of problematic significance, such as the UN Declaration of Human Rights. It is one thing to get the world's states to sign up to such a declaration, but it is another to make it effective in the face of the resolute defence by states of their own sovereignty and right to interpret the meaning of the declaration according to their own lights and cultural norms.

The difficulties alluded to here do not amount to the conclusion that such universally proclaimed commitments are completely empty. But they do suggest that such phenomena represent at best the beginning, rather than the end, of a process of working out and implementing the commitments entered into. A UN Declaration of the Rights of Non-Human Species would be a significant advance, but would clearly be no more than a first, not a last, step on the road to procuring ecological justice. However, as we noted at the start of this chapter, it may even be a virtue of an IEI that it begins life as an apparently empty commitment to a vague formula, provided that those charged with its development use the opportunities that arise to steer the institution in a more effective direction.

One possibility might be the encouragement of the constitutional embedding of protection for non-human species within state constitutions on the basis of state commitments to such a Universal Declaration of the Rights of Non-Human Species. This would commit the signatories to the pursuit of ecological justice within their state boundaries and perhaps also bind them into supporting the work of IEIs which would aim to foster the three Cs across the globe with respect to ecological justice. This would cover monitoring, research into what species exist and what current threats to them are likely to be particularly serious; the establishment of networks of interested actors, such as NGOs, to maintain concerned publics; the provision of funds to poorer states to facilitate their internal ecological justice institutions; and so on. Current efforts to obtain UN recognition of the 'Earth Charter' are an example of this kind of initiative in the environmental field (Callicott 2002: 18–20).

This might lead on in time to the establishment of an International Court of Ecological Justice,[2] although the development of its equivalent in the purely human case is clearly rather problematic, especially given certain states' concerns with protecting their national sovereignty with respect to matters of criminal justice. However, what the analysis of Hass *et al.* strongly suggests is that effective IEIs may be able to achieve their aims by acting as facilitators and monitors of actions taken by states purely internally, and as guarantors of anti-free riding threats with respect to transboundary and open access environmental problems. Success in these respects may obviate the need to look for full-blooded international organizations that look like nascent components of the problematic concept of international government rather than of the more widely accepted concept of international governance.

The example of GRASP

Let us return to the case of the Great Apes Survival Project Partnership for further clues as to what all this might involve. We have already noted the aims and rationale of GRASP, but what is now of interest are the proposed institutional arrangements for achieving those aims. They illustrate the ways in which IEIs can interconnect with each other to foster the three Cs, and the importance of networks involving NGOs, private groups and state governments as well as the IEIs and other international institutions whose purpose is not directly related to GRASP.

The first point to note, of especial significance from the ecological justice perspective, is that the species that are the focus of GRASP are geographically localized within areas which form the territories of twenty three states within the global South, in sub-Saharan Africa and South East Asia. The protection of these species, therefore, will necessarily involve issues of concern and capacity for the governments of some of the poorest states on the planet, governments facing huge, urgent and intractable problems of hunger, poverty and illness within their populations. It would be unsurprising if the agreement of these governments to participate in GRASP was characterized by reluctance and large amounts of lip-service, particularly as the rationale for the project, anthropocentric as it is, does not rest on the generation of large economic benefits among the populations of the states involved. By May 2003 nine of the twenty-three range states had agreed to join GRASP and nominate a member of their government to be the point of contact for GRASP initiatives (GRASP 2003: 4).

One would expect the governments of these states to seek to extract some quid pro quo from the governments of states in the affluent North whose populations contain many of the concerned publics that have given the impetus to GRASP. In other words, the issue of international equity here will be particularly important – that the costs of the project should be borne by those states and populations committed to the project and best able to afford to pay for it, while the responsibilities of the states which are the site of the project is the different one of facilitating and cooperating in its organization, including the development of concern, if it does not yet exist, among sections of the population whose cooperation will be crucial to the success of the project. However, while these problems may be anticipated, it does not follow that at least some of the governments of such states, or at least some elements in their own populations, are not supportive of the general aims of GRASP. As the GRASP website notes: 'Some communities have longstanding traditions which give special protection to primates, including the great apes. GRASP will build on these wherever possible' (GRASP 2003: 6). Concern may be less of a problem than capacity and the allaying of free-rider fears among the affected governments, particularly in the case of transboundary issues between the African states involved.

Turning to the institutional arrangements for GRASP, we may begin by noting that it is the joint concern of two UN institutions, one explicitly environmental in its scope – the UN Environment Programme (UNEP) – and the other less explicitly so – the UN Educational, Scientific and Cultural Organization (UNESCO). The GRASP website suggests:

> The two UN bodies, working together, are well placed to create a successful partnership of range states, international conventions, non-governmental bodies, scientists, zoos, charitable donors and commercial interests in a sustained campaign to protect not only great apes, but also the ecosystems of which they are a part.
>
> (GRASP 2003: 2)

The contribution of UNESCO draws on its responsibility for some existing international institutions which can make a contribution to the aims of GRASP. Its Man and Biosphere Programme encompasses the World Network of Biosphere Reserves and has established a Regional Postgraduate School for Integrated Forest Management in Kinshasa. It oversees the World Heritage Convention, which aims to preserve sites of 'outstanding universal value' (GRASP 2003: 3). The GRASP website claims that biosphere reserves and world heritage sites in several of the African and Asian range states are 'critical sites for the survival of great apes' (ibid.). In addition, UNESCO is preparing another important programme, the Central African World Heritage Forest Initiative, which has the objective of improving the management of 'several unique transboundary clusters of forest in the Congo Basin and to assist the countries in the region in submitting them for World Heritage nomination' (ibid.). A crucial element in the key monitoring aspect of the project is also being provided by UNESCO. Its 'Open Initiative' venture with international space agencies, such as the European Space Agency, will employ satellite monitoring of world heritage sites, of which those which contain gorilla habitat are of particular relevance to GRASP.

The contribution of UNEP is directly to oversee the GRASP project, with the executive director and his deputy giving publicity to the project in the course of their normal interactions with governments. But it also has responsibility for several other institutions of relevance to the project. It is responsible for the conventions on Biological Diversity, on Migratory Species and on Trade in Endangered Species (CITES). In addition, it is responsible for the World Conservation Monitoring Centre which, together with UNEP's Division of Early Warning and Assessment, is supplying valuable data to the project.

GRASP itself is also an institution of some complexity. By May 2003 it contained various subsections. Firstly there is a group comprising three eminent 'ape envoys', a special adviser and a technical support team. The members of this group have been appointed for their range of 'fundraising, scientific, practical and diplomatic experience' (GRASP 2003: 4). The job of this group is to advise UNEP over which projects to select.

Secondly there is an internal UNEP team of twelve or so professionals 'from the divisions dealing with monitoring, policy, communications, fundraising, the regions and environmental conventions' which meets monthly in Nairobi to oversee the project's progress and plan its next steps, in close contact with the 'ape envoys' (GRASP 2003: 4).

Thirdly there is a three-monthly teleconference of 'all the formal partners in the alliance' (including UNESCO, the NGOs and the conventions) (GRASP 2003: 5). Finally there is the GRASP website itself, which is intended to act as a sounding board and point of contact for anyone with an interest in the project.

The overarching aim at this early stage is to construct 'an alliance of stake-holders under UNEP's leadership' (GRASP 2003: 4). This alliance is intended to comprise the range states, NGOs (sixteen are cited in the website as already being partners in GRASP), sympathetic governments in non-range states, and charities and commercial enterprises, particularly those 'which can generate sustainable income from great apes and their habitat' (ibid.: 5).

Then there are more specific aims. The first is education and publicity about the plight of the great apes (press events, publications – such as a *World Atlas of Great Apes* to be published by UNEP/UNESCO – the internet, film and electronic broad-casting). There are missions which are being mounted to the range states in order to 'study the problems, consult governments and other authorities in order to feed back the urgent and longer term requirements for maintaining viable ape populations an assisting local people at the same time' (GRASP 2003: 3). The aim here is clearly that of developing capacity in the range states:

> A first step is the preparation of a National Great Ape Survival Plan by authorities. These NGASPs can then form a bridge to wider planning mechanisms including those under the Convention on Biological Diversity (CBD) and other multilateral agreements to protect cross border populations of great apes.
>
> (GRASP 2003: 3)

An important element in the development of practical measures to save the great apes is the implementation of pilot projects by the NGO stakeholders with the assistance of the governments of the range states. This is where we most clearly see the emergence of the economic dimension to the project: 'The emphasis is on projects which deliver tangible benefits, involving local communities and the private sector' (GRASP 2003: 3).

The crucial effects, for good and bad, of private actors – corporations and individuals – upon the conservation of great apes is recognized fully. GRASP aims to publicize and encourage responsible ecotourism, emphasizing its beneficial possibilities for local people. This is viewed as both giving those people a material stake in the success of the project and helping in the effort to produce the concern without which the project is a non-starter. Timber companies involved in habitat destruction and the facilitation of the wild bushmeat trade are also to be targeted to persuade them to take a more responsible approach to the economic exploitation of great ape habitat. Other corporate actors whose effect on habitat can be crucial have been identified as mining companies in Central Africa and South-East Asia.

All of this effort costs money, of course. The range states themselves, as earlier noted, are not well placed to provide it – as the strategy notes: 'Of the 21 African range states, 15 are Least Developed Countries whose collective GNP amounts to just 0.16 per cent of the world total and where incomes average less than three dollars a day' (GRASP 2003: 6). UNEP itself is to provide $750,000 seedcorn money over the period 2001 to 2005. The project has set itself the aim of generating $25 million by 2005 by its own fund-raising efforts. NGO contributions are to be facilitated by 'streamlined banking arrangements' via the Born Free Foundation (ibid.: 7). Governments in the affluent North are being approached for donations (though only the UK and Norway are mentioned on the website as having offered funds by May 2003). Efforts are being made to seek, and reward, private donations and other contributions to the project.

All this represents the early stages of one effort to preserve from extinction a group of non-human animals which, given their closeness to us, ought to be the most promising contenders for survival, even in a world of such gross inequality and human suffering as our own currently is. Yet its prospects of success may not be very high. One cannot fail to be impressed by the fact that the project exists at all, and that it is ensconced within the highest levels of global governance. But it faces formidable obstacles to success, given the dire economic circumstances of most of the states within which it has to operate and that it is itself funded on a very problematic basis.

Nevertheless, from the point of view of ecological justice the project does suggest some of the ways in which a human effort to protect non-human species from extinction might be mounted in the absence of a global government with a constitutionally embedded commitment to ecological justice. Concern and capacity within the range states are vital, obviously. Although proponents of ecological justice are concerned with the continued existence of non-human organisms with no obvious commercial value or attraction for most human beings, and even of ones which pose threats of various kinds to human beings, it is hard to resist the conclusion that, as with GRASP, some attempt to find an economic rationale for the preservation of species and habitats is going to be an unavoidable part of the strategy for success. Ecotourism may form a key part of such a rationale. Its providers may engage in it for reasons of economic self-interest. But at least from the point of view of ecological justice it has the advantage that it gives them a motivation to think about undamaged ecosystems as objects of value. This sets the stage for the moral dimension to appear to be less remote from people's concerns than it might otherwise be – and certainly does a better job in this regard than do those modes of economic involvement with ecosystems that involve their wholesale destruction.

The complex networks of IEIs, NGOs, governments and private actors which the GRASP project exemplifies also show the ways in which there is scope for developing the agenda, and pursuing the aims, of ecological justice by a variety of routes, rather than by seeking a direct assault on the problem at the highest, global, level. If GRASP succeeds in its aims then the great apes will be protected across their range by a combination of:

1   national protection plans within each range state, adequately financed by a combination of the states' own resources, and transfers of funds from non-range states, NGOs, IEIs and private donors;
2   transboundary conventions negotiated between range states, with the good offices and monitoring services of IEIs playing a crucial role;
3   economic incentives for local populations to participate fully in the national plans, involving the transformation of economic activities into sustainable forms, whether it is ecotourism or some more aggressively exploitative form of economic activity that is in question;
4   a widespread concern for the continued preservation of the great apes, within the population of the range states and more generally, fostered by the educational activities of all the stakeholders;
5   a permanent body of people professionally employed to monitor, research and publicize the problems faced by the project of maintaining the existence of the great apes in their natural habitats and to feed back their findings to all the stakeholders mentioned in the previous points.

If such an eventuality were to materialize, proponents of ecological justice would be justified in believing that their own project, which would in effect represent the moralization of projects such as GRASP, would be well within reach. Thus, although proponents of ecological justice have to begin from where matters currently stand, perhaps enough has been said in this chapter to show that the matter of discovering viable institutions for pursuing ecological justice may begin with an examination of how far existing IEIs may be turned in the direction of ecological justice.

The arguments offered in the earlier chapters of this book are intended to help with this endeavour in the forums of debate at domestic and international level where moral arguments may be expected to cut some ice. There is evidence that when matters of environmental concern are in play such moral debates are rather easier to mount and pursue than elsewhere, even though standardly the form of ethics involved is purely anthropocentric. The crucial move is that moral arguments should be permitted at all, for then the debate is in the right logical shape for the proponents of ecological justice to make a significant contribution to it.

The five outcomes just mentioned, although in practice they would save the great apes, nevertheless contain the serious flaw of tying in that salvation to human advantage exclusively. What proponents of ecological justice would want to be able to claim is that, even if we should lose interest in the great apes, even if ecotourism should begin not to pay, even if we found we could destroy the apes and their habitat with no serious effect on anything else of interest to human beings, we should still not do so, for the great apes, as do all other living entities, have a moral status which means that they are entitled to the concern of moral agents even when we have no further interest in them. We apply this to the case of human beings, and it is the argument of this book that no less is required when it is non-human species we are considering, even though there is no punishment which those species can exact upon us if we fail, or refuse, to do so.

13 Conclusion

This book has been engaged in a series of arguments with opponents who are disposed to challenge vigorously virtually every one of its main contentions. Let us briefly recall the main points of debate.

The basic claim upon which the book rests, namely that organisms other than human beings are morally considerable, is still by no means universally accepted, nor is it ever likely to be. The arguments which have been used to support this view are all open to counter-arguments. Thus, the argument from marginal cases, which tries to show that the characteristics that are supposed to confer moral considerability uniquely upon human beings are not possessed by all humans some of the time, or by some humans all of the time, may be resisted on the basis that moral considerability applies to all human beings in virtue of the character-istics possessed by normal, mature human beings. The claim that at least some, and probably many, other creatures do possess moral-status conferring properties may simply be refused credence, and be pigeon-holed into the category of 'anthropomorphism', thereby losing intellectual respectability. The observable fact that human beings do treat other organisms with care and consideration may be held to show that the caring moral attitude appropriate only to other human beings may 'spill over' into treatment of other things, not that those other things are really worthy of moral regard.

Even among those who are persuaded that organisms other than human beings are morally considerable, there are many who will remain unpersuaded that non-sentient organisms fall into this category. On their view such organisms are better thought of as analogous to human artefacts – complex bits of biochemical mechanism, doubtless vitally important to the health of ecosystems, and thus possessing instrumental value for organisms which do count morally, but lacking any moral status themselves. For, lacking sentience, nothing can count for them. They are thus devoid of interests, and, it is claimed, only beings possessing interests can count morally.

Even among those who have got as far as agreeing that all organisms, even the 'merely living', do possess moral considerability, there will be serious disagree-ments with some of the arguments of this book. Two groups come to mind. Firstly, some, within the ranks of those who will probably be well disposed to the general aims of the book, will be unhappy with the arguments which seek to justify the attachment of different moral weights to different kinds of organism. They will instead hold to the view that human beings and other species should rank as moral equals with each other. They will also be strongly inclined to doubt that a system of justice which gives human beings most weight is likely to afford much protection for the interests of the non-human.

Then there will be many who will be strongly inclined to deny that the moral duties which human beings owe to non-human beings involve that of securing distributive justice with respect to those environmental resources necessary for other organisms to survive and thrive after their kind. The most that can be properly asked, they will say, is that human beings treat other organisms with some basic respect – humanely and with care. But the fact that human beings and other organisms are in competition with each other, and that we have to kill them in order to survive, puts the requirement of justice beyond the limits of what may properly count as appropriate moral treatment of such organisms by human beings. Distributive justice with respect to goods and bads is only sensibly discussed, it is often claimed, in connection with beings capable of at least reciprocity towards each other, and probably only in connection with those capable of entering into (quasi-)contractual relationships with each other.

Even among those who are supportive of the idea of ecological justice, and who hold that it is sensible and important to seek to ensure that all organisms have their fair share of environmental resources, there is often resistance to the idea that such a concept of justice may properly be embedded in the constitutional arrangements of a liberal state. For, it is widely agreed, such a state must remain aloof from supporting any substantive view of the good life, which is where many are strongly inclined to locate the concept of ecological justice. Hence, proponents of ecological justice can at best hope to win backing for their cause only in the course of normal democratic debate within the procedures laid down by the constitutions of liberal societies.

Then there are contentious issues of a broader nature than that in which this book is specifically involved, but which have an important bearing upon its argument. As we saw at the start, there will be many who dispute that it is possible any longer to conduct a moral argument upon universalist lines, seeking to generate general moral principles intended to be addressed to, and persuade, any moral intelligence that cares to contemplate them. Moral thought is inevitably contextualized, it will be argued, and must emerge from within social and cultural contexts, rather than be imposed upon them in the name of a spurious universal form of reason.

There are those, often the same thinkers as those committed to the position outlined in the last paragraph, who argue that the modern state is a political structure which is incapable of delivering justice for human beings, let alone for other life-forms, and which is indeed actively working against such justice and for powerful forces with no interest in such justice. Hence, it will be argued, it is at the very least naïve, and quite probably fatuous, to suppose that any significant environmental goal which runs counter to the interests of those powerful human groups that control the modern state can be achieved within the state form.

This pattern of argument often coexists with scepticism about the very idea of the liberal state and of liberalism. Critique of liberalism is almost as old as liberalism itself, and some recent forms of liberal triumphalism have stimulated renewed attacks upon what are seen as its unjustified pretensions to universal validity. The case for ecological justice developed in this book has relied on liberal approaches to the problem of designing the best form of polity for human beings and other organisms. In many ways those approaches are more convincing than they have ever been, as a variety of traditional rivals to liberalism have run afoul of the drawbacks of their own illiberalism. But there is still a rich area of contention to take account of here.

Then there are those who are deeply sceptical of the ability of any moral argument, however well argued and intellectually watertight, to make a real difference to human affairs. Moral argument, it will be said, inevitably leads to moralizing, to which most people have a deep aversion. It is much more fruitful to find reasons for action drawn from self-interest, and material interests will inevitably be rather prominent in this regard, however good a job we do of moralizing self-interest. If your concern is with securing the continued existence of other species, it will be argued, then follow the existing pathways to at least some success. Show that human beings, particularly future human beings, have a direct prudential interest in the continuation of such species, in the form of agricultural, medical, scientific, recreational, economic, cultural, aesthetic, ecological and other benefits (see, for example, Norton 1991: 227). Do not try to persuade them that such beings are owed moral duties by human beings, especially not something as inherently contentious, even in the human case, as duties of justice.

It has to be admitted that all of these positions contain weighty considerations which a proponent of ecological justice needs to consider and counter. This book contains attempts to do just that with respect to most of them, although for reasons of space it spends little time defending liberalism as a general position. But it is possible to be both a reasonable human being and to adopt one or other of the views just outlined which will form sticking points to the arguments of this book. Is there anything else that can be urged in favour of the line of argument pursued by this book which might boost its plausibility and acceptability in the minds of those not already convinced?

To answer this it may be useful to turn down the road of autobiography for a moment, to see what factors in my own life have perhaps made me more open to the line of argument which I have tried to put forward here. I hasten to add, of course, that nothing that follows is supposed to be a substitute for those arguments. If they do not work as arguments then the conclusions which they seek to establish must be rejected.

Firstly, there is the important fact that I have been interested in the natural world from a very early age. I have always loved just walking among fields and hedgerows, seeking out wild flowers, watching birds and wild animals. I have always wanted to know what I am looking at, what it is called, what its habits are and how it lives. My parents shared such interests to a degree, although neither was highly educated or particularly well informed on such matters. They were happy for me to be interested, but did not put any strong emphasis on my pursuing those activities assiduously. It probably helped that I was an only child, and so did not have any siblings to offer me rival interests to those I had developed myself.

I was also aware from the age at which I began to feel drawn to the natural world that this was not a universal trait. My childhood friends showed very little curiosity in such things. An interest in birds, for example, was likely to be developed exclusively in terms of their usefulness as target practice. I wasn't averse to such pursuits myself, but there were other dimensions to my attitudes which they did not share. I was given, as a result, presents such as books to help identify flowers or birds. One identification book for wild flowers which I still possess must have been given to me when I was about seven. I somehow doubt that there are many boys of that age who would be given, or welcome, such a present. In case this seems to be painting a picture of a bookish, solitary only child with nerdish proclivities I should add that I also had many of the usual childhood interests in football, fighting, and so on and had plenty of friends.

What this is meant to convey is the idea of a basic interest in nature which has remained throughout my life as a fundamental point of connection between me and my world. But I was not a country boy, or raised on a farm. The things concerning nature to which I particularly attended – the books I read, TV programmes I watched – were available to all my contemporaries. I do not know why the natural world had such resonance for me. I imagine it might have something to do with a strong visual aesthetic sense, for I was always interested in painting and drawing, and it is the visual sense which above all ties us to the natural as well as to the human-made world. People with no great interest in the natural world often seem to me to have non-visual dimensions to their aesthetic preferences. They are more moved by music or literature, which can have a connection with natural phenomena, but usually a less direct one.

Do I have reason to be thankful for this nature orientation to my life, which seems to go so far back into my biography? I think I do. For I think it gives me a direct awareness of a crucial feature of human existence to which many other people seem to be at least partly blind. This is the fact of our embeddedness within a living world from which we emerged and with which we must forever remain intimately connected. Not everyone who has to deal with this embeddedness directly in their lives, such as farmers and fisherpeople, seem to notice it directly, or to welcome it if they do. This is another aspect of nature awareness which it is worth bringing out – the sense of exhilaration, of positive value directly experienced.

Is this a religious phenomenon? Well, it is certainly the kind of experience which merits the term 'spiritual'. Spirituality need not be religious, in the sense of positing the existence of some all-important transcendent realm. Human spirituality is primarily the search for meaningfulness, and a sense of connectedness with something else – nature, other people, a movement, a deity – is a prime candidate for generating such meaningfulness. A life with no connections beyond the self is the paradigm of a meaningless life, one might say. But I doubt whether any of the usual diagnoses of environmental concern among Western middle-class people, as the attempt to find meaning in a 'disenchanted' world devoid of traditional religious values, applies to my case. My family was in a pretty vague sense Christian in religious orientation, but not church-going, and strongly inclined to distrust organized religion and the 'unco guid'. I had a fairly strong religious belief until my mid-teens. A developing philosophical sense began to undermine it fairly quickly at that point, and I am now completely non-religious in outlook. But my nature orientation long precedes this change in religious viewpoint, and had rather little to do with religious belief at the time that both coexisted in my psyche.

But this awareness of human connectedness with nature for a long time receded from the forefront of my consciousness as my interests turned in a more exclusively human direction, especially in the realm of social and political matters. My political orientation has been left of centre since my teenage years, and until quite recently I would have said of myself that I was fundamentally socialist, and within that diverse family of doctrines I believed that social democracy was the most defensible version. Marxism always interested me, but never convinced me intellectually. It seems to me now to be a good example of a Lakatosian degenerating research programme (see Chalmers 1978: 81–3). It has some initial plausibility and promise. But as time goes on it does not become any more convincing or able to open up new lines of enquiry, rather than finding ways of restating the initial claims in new forms. In practice it turned into a non-theistic religion, with holy books, a priesthood, heresies, witch-hunts and dogma.

There was, however, in my allegiance to socialism always a sense of dissatisfaction. Socialism is humanistic in its moral orientation. For socialists, whether secular or religious, moral or scientific, only human beings really count. That was a view with which I have never really been comfortable, although my discomfort was not really brought to the forefront of my mind until about a decade ago when, in a spirit of intellectual enquiry (and growing ennui with the traditional conservative/socialist/liberal structure to traditional debates within academic and real-world politics) I finally decided to take a proper look at the ideas of the environmental movement and its theorists. I had always been sympathetic to the concerns of environmentalists, but the humanist bent of my education (for example, no issue of human–non-human relationships ever figured in any of the courses of moral philosophy I took) and political orientation had never given me cause to examine the moral and political dimensions of the issues. They always seemed to be essentially practical matters – how to recycle waste, use energy more efficiently and so on.

What I discovered when I looked into this whole new area of thought (for that is what it is, in spite of the fact that some illustrious predecessors have been found for some of the ideas) was that for the first time a moral and political view was being developed which tried to take full account of all the phenomena that had moral significance. The natural world, which in traditional secular and religious humanism was really no more than a backdrop, now assumed its full significance. It suddenly became crystal clear how much of human moral thought, from all traditions, had simply ignored a whole vital dimension. For the first time it was becoming possible to reorientate much that was valuable within these traditions into a perspective in which the human found its proper place – prominent, but not isolated.

One thing this led me to do was to produce my own version of the general point of view that seemed to be emerging from the efforts of environmentally orientated thinkers, for which the label 'ecologism' had been put forward. This is not the loveliest term that could have been devised, and it has the unfortunate implication that supporters of ecologism have to be labelled 'political ecologists' in order to differentiate them from the scientists who study ecology. However, I would now say of myself that I am a political ecologist. In my understanding of it, ecologism is primarily a moral doctrine and the key element is the concept of ecological justice which this book has been attempting to vindicate.

It will be recollected that this detour through autobiography was in response to the question of what else could be said in support of the concept of ecological justice to commend it to those who are not convinced by the specific arguments put forward to counter the particular objections recalled at the start of this final chapter. What may have emerged here is not itself a further argument. Rather it amounts to a plea to doubters to assure themselves that they have done all they can to be open to the possibility that nature is, and specifically other organisms are, of fundamental moral significance. This is a matter of taking an interest in, and becoming informed about, the lives of other species, even if only in a fairly basic way. Such beings are all around us, and it is not a matter, therefore, of pursuing something recondite and hard to grasp, such as quarks.[1] Some people, like me, may take to this interest in other species for hard-to-fathom reasons lodged deep in their natures. But there is no reason to suppose that most people cannot be brought to at least some recognition that lives other than human ones are lived on this planet, and that if you are concerned about how you are to live, then the impact you can have on these lives too is a matter of legitimate moral concern.

Clearly, this cannot guarantee acceptance for the arguments for ecological justice, or ecologism. But it ought to alert everyone to the possibility of a view of the world within which such moral concerns make sense and to the fact that this point of view at least attempts the most comprehensive moral doctrine it is possible to arrive at, rather than restricting moral considerability to only a subsection of organic life. That in itself ought to make the position argued for in this book – though not necessarily the particular arguments put forward – one of the perennial possibilities for moral thought, to be examined and discussed as a matter of course rather than regarded as a strange aberration.

But the concerns of this book are not, of course, solely theoretical. Those who are concerned that we are poised on the brink of a major extinction episode, caused by human activities, are trying by a variety of means to prevent such an outcome. The aim of a theory of ecological justice must in part be to contribute to this effort. But it is apparent to all that ideas can have a significant effect only when human beings are ready to receive them. Whether they are so ready is, unfortunately, something beyond the power of the authors of ideas to determine. Still, ideas are always worth having – and in a book which is devoted to the development and defence of ideas that is probably the best point with which to end.

Notes

2 The case for social constructivism considered

1 See Gare 1995 for an excellent discussion of postmodernism in its relation to environmental problems.
2 All the quotations in this paragraph are from a personal communication.
3 This is roughly what Walzer characterizes immanentism to be in ethics (Walzer 1983).

4 The restriction of moral status to sentient organisms

1 Singer 1993: chapter 3 is the source of this argument.

5 The moral status of the non-sentient

1 In a personal communication.
2 In a personal communication.

6 The concept of ecological justice: objections and replies

1 In a personal communication addressing an earlier draft of these arguments.
2 I am grateful to Andy Dobson for urging me to consider the idea that issues of ecological justice may be resolvable purely on the basis of considering what each (kind of) organism contributes to the total of environmental benefits. As will be apparent, it is not a view which I am able to accept, but it is an important possibility which merits careful consideration.
3 As Marcel Wissenburg has pointed out to me, Rawls in his *Political Liberalism* allows that his principles of justice can be overruled in certain circumstances – by the 'general conception of justice' in poorer societies, and by principles for a decent but non-democratic society. However, the Rawlsian principles are not overturnable in a 'well-ordered' society, whereas there are no circumstances where the rights of non-person organisms may not properly be overridden.

7 Liberal theories of justice and the non-human

1 However, as noted earlier, Hailwood wants to encompass non-biotic nature in the 'otherness' view, and thus wants the 'respect' he enjoins to apply to entities which cannot have any interests. Arguably, whatever this respect amounts to it is going to be rather problematic to regard it as moral respect, rather than, say, aesthetic.

8 Ecological justice and justice as impartiality

1 From a personal communication.

2 I am indebted to John Horton for this point.
3 I am indebted to Andy Dobson for this point.
4 I am indebted to Tim Hayward for enabling me to see the necessity for this point and for prompting me to work out my response to it.
5 A point which John Horton has strongly pressed in a personal communication.
6 Quoted from a personal communication to me from John Horton.

9 Ecological justice and the non-sentient

1 It should be noted at this juncture, in connection to a point made to me by Wissenburg, that we are here considering examples of the 'merely living' which are fully mature specimens of their species. That is, what is here being discussed should not be thought to apply also to immature specimens of creatures which are more than merely living when they reach maturity. Thus, what is here said does not apply to fertilized ova of any species and, *a fortiori*, does not cover human beings at that stage either. There are, of course, issues about what duties we do owe to such beings. But that is a matter separable from the issue of what duties we owe to those creatures which are constitutionally, so to speak, merely living.

11 Institutional arrangements within states

1 The discussion in this section is heavily indebted to an excellent exploration of the issues in Pickett *et al.* (1997).

12 Institutional arrangements at the global level

1 However, it is an implication of the new ecological paradigm mentioned in the last chapter that any adverse effect on the global environment, however localized it appears to be, is a matter in which all moral actors have a legitimate interest, given the potential harm that may lurk within it for the global ecosphere as a whole.
2 Low and Gleeson (1998: 191) argue for such an idea, comprising a World Environment Council and an International Court of the Environment, in their discussion of institutions for environmental and ecological justice.

13 Conclusion

1 However, it is a matter of some concern that in the UK even some students of ecology who train as conservation managers are showing alarming signs of ignorance concerning the identification of common species in the field (Bowler 2003).

References

Attfield, R. (1991) *The Ethics of Environmental Concern*, 2nd edn, Athens: University of Georgia Press.

Barry, B. (1995) *Justice as Impartiality*, Oxford: Clarendon Press.

—— (1998) 'Something in the disputation not unpleasant', in P. Kelly (ed.) *Impartiality, Neutrality and Justice*, Edinburgh: Edinburgh University Press.

—— (1999) 'Sustainability and intergenerational justice', in A. Dobson (ed.) *Fairness and Futurity*, Oxford: Oxford University Press.

Barry, J. (1999) *Rethinking Green Political Theory*, London, Thousand Oaks, CA, and New Delhi: Sage.

Baxter, B. (1996) 'Ecocentrism and persons', *Environmental Values*, 5: 205–19.

—— (1999) *Ecologism: an introduction*, Edinburgh: Edinburgh University Press; Washington DC: Georgetown University Press (2000).

BBC (2000a) *The State of the Planet*, broadcast on 22 November, presented by Sir David Attenborough.

BBC (2000b) *The State of the Planet*, broadcast on 15 November, presented by Sir David Attenborough.

Becker, L. (1977) *Property Rights*, London: Routledge & Kegan Paul.

Beitz, C. (1979) *Political Theory and International Relations*, Princeton, NJ: Princeton University Press.

Bell, D. (2003) 'Political liberalism and ecological justice', paper presented at the 2nd ECPR (European Consortium for Political Research) Biennial Conference, Marburg, Germany, September.

Bennett, J. (1976) *Linguistic Behaviour*, Cambridge: Cambridge University Press.

Bookchin, M. (1995) *From Urbanization to Cities*, rev. edn, London and New York: Cassell.

Bowler, P. (2003) 'Identity crisis: what's the difference between an otter and a mink? Unsure? Don't ask the new breed of conservationist', *The Guardian*, 3 September.

Callicott, J. (1999) 'Silencing philosophers', *Environmental Values*, 8: 499–516.

—— (2002) 'The pragmatic power and promise of theoretical environmental ethics', *Environmental Values*, 11: 3–25.

Cantori, L. (2003) 'Democracy from within Islam', *The CSD [Centre for the Study of Democracy] Bulletin*, 10/2: 1–2.

Carter, A. (1999) *A Radical Green Political Theory*, London: Routledge.

Chalmers, A. (1978) *What is This Thing Called Science?* Milton Keynes: Open University Press.

Christensen, N. (1997) 'Managing for heterogeneity and complexity in dynamic land-scapes', in S. Pickett, R. Ostfeld, M. Shachak and G. Likens (eds) *The Ecological Basis for Conservation: heterogeneity, ecosystems and biodiversity*, New York: Chapman & Hall.

Connolly, K. (2002) 'German animals given legal rights', *The Guardian*, 22 June.

Daly, H. and Cobb, J. (1990) *For the Common Good*, London: Green Print.

DeGrazia, D. (1996) *Taking Animals Seriously: mental life and moral status*, Cambridge: Cambridge University Press.

De Waal, F. (1998) *Chimpanzee Politics: power and sex among apes*, rev. edn, Baltimore, MA: Johns Hopkins University Press.

Diamond, J. (1992) *The Rise and Fall of the Third Chimpanzee*, London: Vintage.

Dobson, A. (1996) 'Representative democracy and the environment', in W. Lafferty and J. Meadowcroft (eds) *Democracy and the Environment*, Cheltenham and Brookfield, VT: Edward Elgar.

—— (1998) *Justice and the Environment*, Oxford: Oxford University Press.

—— (ed.) (1999) *Fairness and Futurity*, Oxford: Oxford University Press.

—— (2000) 'Sustainable development and the defence of the natural world', in K. Lee, A. Holland and D. McNeill (eds) *Global Sustainable Development in the Twenty-First Century*, Edinburgh: Edinburgh University Press.

—— (2004) 'Social inclusion, environmental sustainability and citizenship education', in J. Barry, B. Baxter and R. Dunphy (eds) *Europe, Globalization and Sustainability*, London: Routledge.

Dowie, M. (1995), *Losing Ground: American environmentalism at the close of the twentieth century*, Cambridge, MA and London: MIT Press.

Doyle, T. and McEachern, D. (1998) *The Environment and Politics*, London and New York: Routledge.

Forest Peoples Project (2003) 'Indigenous peoples and protected areas in Africa: from principles to practice', online: http://forestpeoples.gn.apc.org/Briefings/Africa/fpproj_update_mar03_eng.htm (accessed 4 November 2003).

Gare, A. (1995) *Postmodernism and the Environmental Crisis*, London and New York: Routledge.

Goodin, R. (1992) *Green Political Theory*, Cambridge: Polity.

Goodpaster, K. (1978) 'On being morally considerable', *Journal of Philosophy*, 75: 308–25.

GRASP (2003) online: http://www.unep.org/grasp/ (accessed 21 May 2003).

Gutman, A. (1995) 'Justice across the spheres', in D. Miller and M. Walzer (eds) *Pluralism, Justice and Equality*, Oxford: Oxford University Press.

Haas, P., Keohane, R. and Levy, M. (eds) (1993) *Institutions for the Earth: sources of effective international environmental protection*, Cambridge, MA and London: MIT Press.

Habermas, J. (1990) *Moral Consciousness and Communicative Action*, trans. C. Lenhardt and S. Nicholson, Cambridge: Polity.

Hailwood, S. (2003) 'A new green liberalism', paper presented at the 2nd ECPR (European Consortium for Political Research) Biennial Conference, Marburg, Germany, September.

Harris, P. (2001) *International Equity and Global Environmental Politics*, Aldershot: Ashgate.

Hayek, F. (1960) *The Constitution of Liberty*, London and Henley: Routledge & Kegan Paul.

Hayward, T. (1998) *Political Theory and Ecological Values*, Cambridge: Polity.

Hegel, G. (1952) *The Philosophy of Right*, trans. T. Knox, Oxford: Clarendon Press.

Hurrell, A. and Kingsbury, B. (eds) (1992) *The International Politics of the Environment*, New York: Oxford University Press.

Johnson, L (1993) *A Morally Deep World*, Cambridge: Cambridge University Press.

Keesing's Record of World Events, 45 (1999), March.

Kelsey, E. (2003) 'Integrating multiple knowledge systems into environmental decision-making: two case studies of participatory biodiversity initiatives in Canada and their implications for conceptions of education and public involvement', *Environmental Values*, 12: 381–96.

Korsgaard, C. (1996) *The Sources of Normativity*, Cambridge: Cambridge University Press.

Kuhn, T. (1970) *The Structure of Scientific Revolutions*, 2nd edn, Chicago, IL: University of Chicago Press.

Kuhse, H. (ed.) (2002) *Peter Singer: unsanctifying human life*, Oxford and Malden, MA: Blackwell.

Leakey, R. and Lewin, R. (1996) *The Sixth Extinction*, London: Phoenix.

Le Prestre, P. (2002a) 'The long road to a new order', in Le Prestre (ed.) *Governing Global Biodiversity: the evolution and implementation of the Convention on Biological Diversity*, Aldershot: Ashgate.

Le Prestre, P. (ed.) (2002b) *Governing Global Diversity: the evolution and implementation of the Convention on Biological Diversity*, Aldershot: Ashgate.

Lewis, D. (1969) *Convention: a philosophical study*, Cambridge, MA: Harvard University Press.

Lovelock, J. (1979) *Gaia: a new look at life on earth*, Oxford: Oxford University Press.

Low, N. and Gleeson, B. (1998) *Justice, Society and Nature: an exploration of political ecology*, London: Routledge.

McFarland, D. (ed.) (1981) *The Oxford Companion to Animal Behaviour*, Oxford: Oxford University Press.

McGraw, D. (2002) 'The story of the Biodiversity Convention: from negotiation to implementation', in P. Le Prestre (ed.) *Governing Global Biodiversity*, Aldershot: Ashgate.

MacIntyre, A. (1988) *Whose Justice? Which Rationality?*, London: Duckworth.

Meyer, J. (1997) 'Conserving ecosystem function', in S. Pickett, R. Ostfeld, M. Shachak and G. Likens (eds) *The Ecological Basis of Conservation: heterogeneity, ecosystems, and biodiversity*, New York: Chapman & Hall.

Midgely, M. (1983) *Animals and Why They Matter*, Athens: University of Georgia Press.

Millennium Ecosystem Assessment (2003) online: http://www.millenniumassessment.org/en/news/ (accessed 18 August 2003).

Nagel, T. (1991) *Equality and Partiality*, Oxford: Oxford University Press.

Neanderthalers and Modern Humans: a regional guide (2004) online: http://www.neanderthal-modern.com/ (accessed 7 January 2004).

Norton, B. (1991) *Toward Unity among Environmentalists*, Oxford and New York: Oxford University Press.

Nozick, R. (1974) *Anarchy, State and Utopia*, Oxford: Blackwell.

O'Neill, G. and Attiwill, P. (1997) 'Getting ecological paradigms into the political debate: or will the messenger be shot?', in S. Pickett, R. Ostfeld, M. Shachak and G. Likens (eds) *The Ecological Basis of Conservation: heterogeneity, ecosystems, and biodiversity*, New York: Chapman & Hall.

O'Neill, J. (1993) *Ecology, Policy and Politics: human well-being and the natural world*, London: Routledge.

Orwell, G. (1971) *1984*, Harmondsworth: Penguin.

Peace Parks Foundation (2003) online: http://www.peaceparks.org/ (accessed 17 November 2003).

Perlo, K. (2003) 'Animal truth: the role of the animal in the development of human world views', unpublished PhD thesis, University of Dundee.

Pickett, S., Ostfeld, R., Shachak, M. and Likens, G. (eds) (1997) *The Ecological Basis of Conservation: heterogeneity, ecosystems, and biodiversity*, New York: Chapman & Hall.

Pinker, S. (2002) *The Blank Slate: the modern denial of human nature*, London: Allen Lane.

Porter, G. and Brown, J. W. (1996) *Global Environmental Politics*, 2nd edn, Boulder, CO: Westview Press.

Possingham, H. (1997) 'State-dependent decision analysis for conservation biology', in

S. Pickett, R. Ostfeld, M. Shachak and G. Likens (eds) *The Ecological Basis of Conservation: heterogeneity, ecosystems, and biodiversity*, New York: Chapman & Hall.

Potvin, C., Revéret, J.-P., Patenaude, G. and Hutton, J. (2002) 'The role of indigenous peoples in conservation actions: a case study of cultural differences and conservation priorities', in P. Le Prestre (ed.) *Governing Global Biodiversity: the evolution and implementation of the Convention on Biological Diversity*, Aldershot: Ashgate.

Preston, C. (2002) 'Animality and morality', *Environmental Values*, 11: 427–42.

Pulliam, H. (1997) 'Providing the scientific information that conservation practitioners need', in S. Pickett, R. Ostfeld, M. Shachak and G. Likens (eds) *The Ecological Basis of Conservation: heterogeneity, ecosystems, and biodiversity*, New York: Chapman & Hall.

Radford, T. (2003a) 'Goodbye cruel world', *The Guardian* ('Life' supplement) 2 October: 4–6.

—— (2003b) 'Injustice makes monkeys go cool on cucumbers', *The Guardian*, 19 September: 9.

Rawls, J. (1972) *A Theory of Justice*, Oxford: Oxford University Press.

—— (1993) *Political Liberalism*, New York: Columbia University Press.

—— (1999) *A Theory of Justice*, rev. edn, Oxford: Oxford University Press.

—— (2001) *Justice as Fairness: a restatement*, Cambridge, MA and London: Belknap Press.

Regan, T. (1983) *The Case for Animal Rights*, Berkeley, CA: University of California Press.

Rogers, K. (1997) 'Operationalizing ecology under a new paradigm: an African perspective', in S. Pickett, R. Ostfeld, M. Shachak and G. Likens (eds) *The Ecological Basis of Conservation: heterogeneity, ecosystems, and biodiversity*, New York: Chapman & Hall.

Rolston, H. (1997) 'Nature for real: is nature a social construct?', in T. Chappell (ed.) *The Philosophy of the Environment*, Edinburgh: Edinburgh University Press.

Rorty, R. (1980) *Philosophy and the Mirror of Nature*, Princeton, NJ: Princeton University Press.

Sachs, W. (ed.) (1993) *Global Ecology: a new arena of political conflict*, London and Atlantic Highlands, NJ: Zed Books.

Scottish Natural Heritage (1999) *Scotland's Wildlife: bats and people*, Perth: SNH.

Selby-Bigge, L. (ed) (1888) *A Treatise of Human Nature by David Hume*, Oxford: Clarendon Press.

—— (ed.) (1902) *Enquiries Concerning the Human Understanding and Concerning the Principles of Morals by David Hume*, 2nd edn, Oxford: Clarendon Press.

Shiva, V. (1991) *The Violence of the Green Revolution: Third World agriculture, ecology and politics*, London and New York: Zed Books.

Sikora, R. and Barry, B. (eds) (1996) *Obligations to Future Generations*, Cambridge: White Horse Press.

Singer, P. (1990) *Animal Liberation*, 2nd edn, New York: New York Review.

—— (1993) *Practical Ethics*, 2nd edn, Cambridge: Cambridge University Press.

—— (1999) *A Darwinian Left: politics, evolution and cooperation*, London: Weidenfeld & Nicolson.

—— (2003) 'Animal liberation at 30', *New York Review of Books*, 15 May: 23–6.

Smart, J. and Williams, B. (1973) *Utilitarianism: for and against*, Cambridge: Cambridge University Press.

Smith, M. (2001) *An Ethics of Place: radical ecology, postmodernity and social theory*, Albany, NY: SUNY Press.

Sylvan, R. and Bennett, D. (1994) *The Greening of Ethics: from anthropocentrism to deep-green theory*, Cambridge: White Horse Press; Tucson, AR: University of Arizona Press.

Szasz, A. (1994) *Ecopopulism: toxic waste and the movement for environmental justice*, Minneapolis and London: University of Minneapolis Press.

Van Lawick-Goodall, J. (1971) *In the Shadow of Man*, Glasgow: William Collins.

Walzer, M. (1983) *Spheres of Justice: a defence of pluralism and equality*, Oxford: Blackwell.

Warren, K. (2000) *Ecofeminist Philosophy: a Western perspective on what it is and why it matters*, Lanham, MD: Rowman & Littlefield.

Warren, M. (1997) *Moral Status: obligations to persons and other living things*, Oxford: Clarendon Press.

Weale, A. (1998) 'From contracts to pluralism?', in P. Kelly (ed.) *Impartiality, Neutrality and Justice*, Edinburgh: Edinburgh University Press.

Wetlesen, J. (1999) 'The moral status of beings who are not persons: a casuistic argument', *Environmental Values*, 8: 287–323.

Wiens, J. (1997) 'The emerging role of patchiness in conservation biology', in S. Pickett, R. Ostfeld, M. Shachak and G. Likens (eds) *The Ecological Basis of Conservation: heterogeneity, ecosystems, and biodiversity*, New York: Chapman & Hall.

Williams, B. (1981) *Moral Luck: philosophical papers 1973–1980*, Cambridge: Cambridge University Press.

Wilson, E. (1984) *Biophilia*, Cambridge, MA: Harvard University Press.

—— (1994) *The Diversity of Life*, Harmondsworth: Penguin.

Wissenburg, M. (1993) 'The idea of nature and the nature of distributive justice', in A. Dobson and P. Lucardie (eds) *The Politics of Nature: explorations in green political theory*, London: Routledge.

—— (1998) *Green Liberalism: the free and green society*, London: UCL Press.

Index